Hirotaka Fujimoto

Value Distribution Theory of the Gauss Map of Minimal Surfaces in $\mathbf{R}^m$

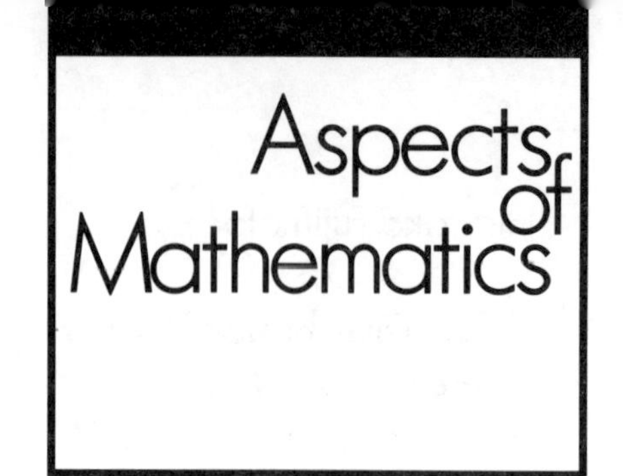

Edited by Klas Diederich

Vol. E 1: G. Hector/U. Hirsch: Introduction to the Geometry of Foliations, Part A
Vol. E 2: M. Knebusch/M. Kolster: Wittrings
Vol. E 3: G. Hector/U. Hirsch: Introduction to the Geometry of Foliations, Part B
Vol. E 4: M. Laska: Elliptic Curves of Number Fields with Prescribed Reduction Type (out of print)
Vol. E 5: P. Stiller: Automorphic Forms and the Picard Number of an Elliptic Surface
Vol. E 6: G. Faltings/G. Wüstholz et al.: Rational Points*
Vol. E 7: W. Stoll: Value Distribution Theory for Meromorphic Maps
Vol. E 8: W. von Wahl: The Equations of Navier-Stokes and Abstract Parabolic Equations (out of print)
Vol. E 9: A. Howard/P.-M. Wong (Eds.): Contributions to Several Complex Variables
Vol. E 10: A. J. Tromba (Ed.): Seminar of New Results in Nonlinear Partial Differential Equations*
Vol. E 11: M. Yoshida: Fuchsian Differential Equations*
Vol. E 12: R. Kulkarni, U. Pinkall (Eds.): Conformal Geometry*
Vol. E 13: Y. André: G-Functions and Geometry*
Vol. E 14: U. Cegrell: Capacities in Complex Analysis
Vol. E 15: J.-P. Serre: Lectures on the Mordell-Weil Theorem
Vol. E 16: K. Iwasaki/H. Kimura/S. Shimomura/M. Yoshida: From Gauss to Painlevé
Vol. E 17: K. Diederich (Ed.): Complex Analysis
Vol. E18: W. W. J. Hulsbergen: Conjectures in Arithmetic Algebraic Geometry
Vol. E19: R. Racke: Lectures on Nonlinear Evolution Equations
Vol. E20: F. Hirzebruch, Th. Berger, R. Jung: Manifolds and Modular Forms*
Vol. E21: H. Fujimoto: Value Distribution Theory of the Gauss Map of Minimal Surfaces in $\mathbf{R}^m$

*A Publication of the Max-Planck-Institut für Mathematik, Bonn

Volumes of the German-language subseries "Aspekte der Mathematik" are listed at the end of the book.

Hirotaka Fujimoto

Value Distribution Theory of the Gauss Map of Minimal Surfaces in $\mathbf{R}^m$

Die Deutsche Bibliothek – CIP-Einheitsaufnahme

Fujimoto, Hirotaka: Value distribution of thc
Gauss map of minimal surfaces in Rm /
Hirotaka Fujimoto. – Braunschweig; Wiesbaden:
Vieweg, 1993
(Aspects of mathematics: E; Vol. 21)
ISBN 3-528-06467-6

NE: Aspects of mathematics / E

Professor *Hirotaka Fujimoto*
Department of Mathematics Faculty of Science
Kanazawa University
Marunouchi, Kanazawa, 920
Japan

Mathematics Subject Classification: 53-02, 53A10, 30-02, 30D35

Cover design: Wolfgang Nieger, Wiesbaden
Printing and binding: Lengericher Handelsdruckerei, Lengerich
Printed on acid-free paper
Printed in Germany

ISSN 0179-2156

ISBN 3-528-06467-6

Preface

These notes are based on lectures given at Instituto de Matematica e Estatistica in Universidade de São Paulo in the spring of 1990. The main subject is function-theoretic, particularly, value-distribution-theoretic properties of the Gauss maps of minimal surfaces in the euclidean space.

The classical Bernstein theorem asserts that there is no nonflat minimal surface in $\mathbf{R}^3$ which is described as the graph of a C^2-function on $\mathbf{R}^2$. On the other hand, the classical Liouville theorem asserts that there is no bounded nonconstant holomorphic function on the complex plane. The conclusions of these theorems have a strong resemblance. Bernstein's theorem was improved by many reseachers in the field of differential geometry, Heinz, Hopf, Nitsche, Osserman, Chern and others. On the other hand, in the field of function theory, Liouville's theorem was improved as the Casoratti-Weierstrass theorem, Picard's theorem and Nevanlinna theory, which were generalized to the case of holomorphic curves in the projective space by E. Borel, H. Cartan, J. and H. Weyl and L. V. Ahlfors. These results in two different fields are closely related. As for recent results, in 1988 the author gave a Picard type theorem for the Gauss map of minimal surfaces which asserts that the Gauss map of nonflat complete minimal surfaces in $\mathbf{R}^3$ can omit at most four values. Moreover, he obtained modified defect relations for the Gauss map of complete minimal surfaces in $\mathbf{R}^m$, which have analogies to the defect relation in Nevanlinna theory. Moreover, several results related to these subjects were obtained by X. Mo-R. Osserman, S. J. Kao, M. Ru and so on.

In this book, after developing the classical value distribution theory of holomorphic curves in the projective space, we will explain the above-mentioned modified defect relations for the Gauss map of complete minimal surfaces together with detailed proofs and related results.

As for prerequisites, it is assumed that the reader is acquainted with basic notions in function theory and differential geometry. Although the description is intended to be self-contained as far as possible, he will need some basic definitions and results concerning exterior algebra, differentiable manifold, Riemann surface and projective space, all of which, except for a few theorems, appear in standard texts for graduate students.

The author would like to express his thanks to Plinio Amarante Quirio Simões for inviting him to IME, for giving him the opportunity to lecture and for his valuable comments. In addition, the author would like to express his thanks to Antonio Carlos Aspert and Dr. Bennett Palmer for reading the manuscript and pointing out some mistakes; and to T. Osawa and K. Diederich for recommending to him that he writes this book.

Kanazawa, Japan
November 1992

Hirotaka FUJIMOTO

Table of Contents

Introduction

Let $x = (x_1, x_2, x_3) : M \to \mathbf{R}^3$ be an oriented surface immersed in $\mathbf{R}^3$. By definition, the classical Gauss map $g : M \to S^2$ is the map which maps each point $p \in M$ to the point in S^2 corresponding to the unit normal vector of M at p. On the other hand, S^2 is canonically identified with the extended complex plane $\mathbf{C} \cup \{\infty\}$ or $P^1(\mathbf{C})$ by the stereographic projection. We may consider the Gauss map g as a map of M into $P^1(\mathbf{C})$. Using systems of isothermal coordinates, we can consider M as a Riemann surface. Our main concerns are minimal surfaces, namely, surfaces which have minimal areas for all small perturbations. In 1915, S. Bernstein gave the following theorem([5]):

THEOREM 1. *If a minimal surface M in $\mathbf{R}^3$ is given as the graph*

$$M := \{(x_1, x_2, f(x_1, x_2)); (x_1, x_2) \in \mathbf{R}^2\}$$

of a C^2-function $f(x_1, x_2)$ on (x_1, x_2)-plane, then M is necessarily a plane.

Thirty-seven years later, E. Heinz obtained the following improvements of this([45]) and E. Hopf and J. C. C. Nitche gave some related results([48], [56]).

THEOREM 2. *If a minimal surface M is given as the graph of a function of class C^2 on a disk $\Delta_R := \{(x_1, x_2); x_1^2 + x_2^2 < R^2\}$ in the (x_1, x_2)-plane, then there is a positive constant C, not depending on M, such that*

$$|K(0)| \leq C/R^2$$

holds, where $K(0)$ is the Gaussian curvature of M at the origin.

Later, R. Osserman generalized these results to a minimal surface which need not to be the graph of a function ([59], [60]). One of his results is stated as follows:

THEOREM 3 ([59]). *Let M be a simply-connected minimal surface immersed in $\mathbf{R}^3$ and assume that there is some fixed nonzero vector n_0 and a number $\theta_0 > 0$ such that all normals to M make angles at least θ_0 with n_0. Then*

$$|K(p)|^{1/2} \leq \frac{1}{d(p)} \frac{2 \cos(\theta_0/2)}{\sin^3(\theta_0/2)} \qquad (p \in M),$$

where $d(p)$ denotes the distance from p to the boundary of M.

Moreover, R. Osserman proved the following fact in the paper [61].

THEOREM 4. *Let $x : M \to \mathbf{R}^3$ be a nonflat complete minimal surface immersed in $\mathbf{R}^3$. Then the complement of the image of the Gauss map is of logarithmic capacity zero in $P^1(\mathbf{C})$.*

Theorem 4 is an improvement of Theorem 1. In fact, under the assumption of Theorem 1 the Gauss map of M omits the set which corresponds to the lower half of the unit sphere, which is not be of logarithmic capacity zero. Therefore, M is necessarily flat and consequently a plane.

In 1981, Theorem 4 was improved by F. Xavier in his paper [76] as follows.

THEOREM 5. *In the same situation as in Theorem 4, the Gauss map of M can omit at most six points in $P^1(\mathbf{C})$.*

Afterwards, the author obtained the following theorem ([36]).

THEOREM 6. *The number of exceptional values of the Gauss map of a nonflat complete minimal surface immersed in $\mathbf{R}^3$ is at most four.*

Here, the number four is the best-possible. In fact, there are many examples of nonflat complete minimal surfaces immersed in $\mathbf{R}^3$ whose Gauss maps omit exactly four values([63]). Among them, Scherk's surface is the most famous.

Recently, the author gave also the following estimate of the Gaussian curvature of a minimal surface in $\mathbf{R}^3$ related to Theorem 3([41]).

THEOREM 7. *Let $x = (x_1, x_2, x_3) : M \to \mathbf{R}^3$ be a minimal surface immersed in $\mathbf{R}^3$ and let $G : M \to S^2$ be the Gauss map of M. Assume that G omits five distinct unit vectors $n_1, \ldots, n_5 \in S^2$. Let θ_{ij} be the angle between n_i and n_j and set*

$$L := \min\left\{\sin\left(\frac{\theta_{ij}}{2}\right); 1 \leq i < j \leq 5\right\}.$$

Then, there exists some positive constant C, not depending on each minimal surface, such that

$$|K(p)|^{1/2} \leq \frac{C}{d(p)} \frac{\log^2 \dfrac{1}{L}}{L^3} \qquad (p \in M).$$

Here, by presenting an example we can show that the exponent of L in the denominator of the right hand side of the above inequality cannot be replaced by any number smaller than three.

The above-mentioned results are closely related to value distribution theory of meromorphic functions on the complex plane. Theorem 1 is related to Liouville's theorem and Theorem 2 is something like the classical Landau's theorem. Moreover, Theorem 6 is very similar to the classical Picard theorem for meromorphic functions on $\mathbf{C}$.

We can give another analogy between the value distribution of holomorphic maps of $\mathbf{C}$ into $P^n(\mathbf{C})$ and that of the Gauss map of complete minimal surfaces in $\mathbf{R}^m$. To state this, we first explain the Gauss map of an oriented surface immersed in $\mathbf{R}^m$. We consider the set Π of all oriented 2-planes in $\mathbf{R}^m$ containing the origin. For each $P \in \Pi$ we take a positively oriented orthonormal basis (X, Y) of P and to P we correspond the point

$$\Phi(P) := \pi(X - \sqrt{-1}Y) \in P^{m-1}(\mathbf{C}),$$

where $\pi : \mathbf{C}^m - \{0\} \to P^{m-1}(\mathbf{C})$ denotes the canonical projection. It is easily seen that $\Phi(P)$ does not depend on a choice of orthonormal basis. Moreover, we can show that Φ is a bijection of Π onto the quadric

$$Q_{m-2}(\mathbf{C}) := \{(w_1 : \cdots : w_m); w_1^2 + \cdots + w_m^2 = 0\}$$

in $P^{m-1}(\mathbf{C})$. The Gauss map G of an oriented surface $x : M \to \mathbf{R}^m$ is defined as the map which maps each point $p \in M$ to the point $G(p) := \Phi(T_p(M))$, where $T_p(M)$ denotes the oriented tangent plane of M at p translated by the parallel translation which maps p to the origin.

We can prove the following theorem.

THEOREM 8. *Let M be a nonflat complete minimal surface immersed in $\mathbf{R}^m$ and G the Gauss map of M which is considered as a map into $P^{m-1}(\mathbf{C})$. If G omits q hyperplanes in $P^{m-1}(\mathbf{C})$ located in general position, then we have necessarily*

$$q \leq \frac{m(m+1)}{2}.$$

This is an analogy of the following restatement of Borel's theorem([6]), which is a generalization of Picard's theorem to holomorphic curves in $P^n(\mathbf{C})$.

THEOREM 9. *Let $f : \mathbf{C} \to P^n(\mathbf{C})$ be a holomorphic map which is nondegenerate, namely, whose image is not included in any hyperplane in $P^n(\mathbf{C})$. If f omits q hyperplanes in general position, then we have necessarily $q \leq n+1$.*

R. Nevanlinna introduced a new notion of defect and obtained the so-called defect relation for a nonconstant meromorphic function on $\mathbf{C}$ which is a refinement of Picard's theorem([55]). Afterwards, his theory was generalized to the case of holomorphic maps into $P^n(\mathbf{C})$ by H. Cartan([7]), L. Ahlfors([1]) and H. and J. Weyl([73]). Roughly speaking, for a nondegenerate holomorphic map $f : \mathbf{C} \to P^n(\mathbf{C})$ and a hyperplane H the defect $\delta_f(H)$ of H for f is a quantity which represents how small the set $f^{-1}(H)$ is in the complex plane. It always holds that $0 \leq \delta_f(H) \leq 1$ and, for example, if $f^{-1}(H) = \emptyset$, then we have $\delta_f(H) = 1$.

The defect relation for holomorphic curves in $P^n(\mathbf{C})$ given by H. Cartan and others is stated as follows.

THEOREM 10. *Let $f : \mathbf{C} \to P^n(\mathbf{C})$ be a nondegenerate holomorphic map. Then, for arbitrary hyperplanes $H_1, \ldots, H_q$ in general position,*

$$\sum_{j=1}^{q} \delta_f(H_j) \leq n+1.$$

Theorem 10 is a refinement of Theorem 9. In fact, this implies that there are at most $n+1$ hyperplanes H in general position such that $\delta_f(H) = 1$.

In this book, for a nonconstant holomorphic map f of an open Riemann surface with a conformal metric into $P^n(\mathbf{C})$, we define the modified defect $D_f(H)$ of H, which has the properties analogous to the classical defect, and give some modified defect relations. As an application, we shall give the following modified defect relation for the Gauss map of complete minimal surfaces.

THEOREM 11. *Let M be a nonflat complete minimal surface immersed in $\mathbf{R}^m$ with infinite total curvature. Then, for the Gauss map G of M and arbitrary hyperplanes $H_1, \ldots, H_q$ in $P^{m-1}(\mathbf{C})$ located in general position, it holds that*

$$\sum_{j=1}^{q} D_G(H_j) \leq \frac{m(m+1)}{2}.$$

This implies the above-mentioned Theorem 8. We can also prove more precise modified defect relations for the Gauss map of complete minimal surfaces in $\mathbf{R}^3$, which implies Theorem 6, and in $\mathbf{R}^4$.

Chapter 1 deals largely with the Gauss map of minimal surfaces in $\mathbf{R}^3$. The first two sections are expended to prepare some basic facts about surfaces in $\mathbf{R}^m$ for later use, particularly, the Gauss map and Gaussian curvature for minimal surfaces in $\mathbf{R}^m$, which are described along the line of [63]. In §1.3, we explain the Enneper-Weierstrass representation formula for minimal surfaces and, in §1.4, give the so-called sum to product estimates for meromorphic functions. In §1.5, we give a new proof of the big Picard theorem and discuss some related properties of meromorphic functions by the use of sum to product estimates. After these preparations, we prove a refinement of Theorem 7 in §1.6.

In Chapter 2, after explaining some basic facts about holomorphic curves in $P^n(\mathbf{C})$ and their derived curves, which are called associated curves in [73], we give some properties of contact functions and prove sum to product estimates for derived curves. We also explain contracted curves for the derived curves in §2.6 for later use.

Chapter 3 is devoted to the exposition of the value distribution theory of holomorphic curves in $P^n(\mathbf{C})$. We describe the first and second main theorems in the first two sections and the classical defect relation for holomorphic curves in §3.3, which includes Nochka's result on Cartan's conjecture. The proofs of these are given along the lines of [13], which we may call the negative curvature method. In §3.4, we give some applications of the defect relation and, after giving some preparations in §3.5, we prove the defect relation for the derived curves of a holomorphic curve in the last section of this chapter.

In Chapter 4, after some preliminary discussions, we introduce the modified defect and give a modified defect relation for holomorphic curves in Osgood space $P^{n_1}(\mathbf{C}) \times \cdots \times P^{n_L}(\mathbf{C})$.

In the last chapter, we discuss the value distribution of the Gauss map of minimal surfaces in $\mathbf{R}^m$. In the first two sections, we describe the results given by R. Osserman and S. S. Chern concerning complete minimal surfaces of finite total curvature([62], [12]). In §5.3, we prove modified defects relations for the Gauss maps of complete minimal surfaces and, in §5.4, a more precise form of the modified defect relations for the classical Gauss maps of complete minimal surfaces in $\mathbf{R}^3$ and $\mathbf{R}^4$. In the last section, we give some examples which show that the number $m(m+1)/2$ in Theorem 8 is the best-possible for an arbitrary odd number $m(\geq 3)$.

Chapter 1

The Gauss map of minimal surfaces in $\mathbf{R}^3$

§1.1 Minimal surfaces in $\mathbf{R}^m$

In this section we investigate some basic facts in differential geometry. We begin by recalling some notions concerning a surface

$$x = (x_1, \dots, x_m) : M \to \mathbf{R}^m$$

immersed in $\mathbf{R}^m$, which means that M is a connected, oriented real 2-dimensional differentiable manifold and x is a differentiable map of M into $\mathbf{R}^m$ which has maximal rank everywhere.

For a point $p \in M$, take a system of local coordinates (u_1, u_2) around p which are positively oriented. The vectors $\partial x/\partial u_1$ and $\partial x/\partial u_2$ are tangent to M at p and linearly independent because x is an immersion. This shows that the tangent plane of M at p is given by

$$T_p(M) := \left\{ \lambda \frac{\partial x}{\partial u_1} + \mu \frac{\partial x}{\partial u_2} \; ; \; \lambda, \mu \in \mathbf{R} \right\}$$

and the space of all vectors which are normal to M at p, say the normal space of M at p, is given by

$$N_p(M) := \left\{ N; \left(N, \frac{\partial x}{\partial u_1} \right) = \left(N, \frac{\partial x}{\partial u_2} \right) = 0 \right\},$$

where (X, Y) denotes the inner product of vectors X and Y. The metric ds^2 on M induced from the standard metric on $\mathbf{R}^m$, called *the first fundamental form* on M, is given by

$$\begin{aligned} ds^2 = &|dx|^2 := (dx, dx) \\ &= \left(\frac{\partial x}{\partial u_1} du_1 + \frac{\partial x}{\partial u_2} du_2, \frac{\partial x}{\partial u_1} du_1 + \frac{\partial x}{\partial u_2} du_2 \right) \\ &= g_{11} du_1^2 + 2 g_{12} du_1 du_2 + g_{22} du_2^2, \end{aligned}$$

where

$$(1.1.1) \qquad g_{ij} := \left(\frac{\partial x}{\partial u_i}, \frac{\partial x}{\partial u_j} \right) \qquad (1 \leq i, j \leq 2).$$

Take an arbitrary (smooth) regular curve $\gamma : x = x(t)$ $(a < t < b)$ with $x(0) = p$ on M, where $a < 0 < b$. For local coordinates u_1, u_2 around p given by a diffeomorphism Φ of an open neighborhood of p onto an open set in $\mathbf{R}^2$, we may represent γ as

$$\gamma : u_i = u_i(t) \qquad (i = 1, 2)$$

by the use of the functions $u_1(t), u_2(t)$ with $(\Phi \circ x)(t) = (u_1(t), u_2(t))$. For each $N \in N_p(M)$ we set

$$b_{ij}(N) := \left(\frac{\partial^2 x}{\partial u_i \partial u_j}, N \right) \qquad (1 \leq i, \ j \leq 2).$$

For another local coordinates $\tilde{u}_1, \tilde{u}_2$, we have

$$\begin{aligned} \tilde{b}_{k\ell}(N) &:= \left(\frac{\partial^2 x}{\partial \tilde{u}_k \tilde{u}_\ell}, N \right) \\ &= \sum_{i,j} \left(\frac{\partial^2 x}{\partial u_i u_j}, N \right) \frac{\partial u_i}{\partial \tilde{u}_k} \frac{\partial u_j}{\partial \tilde{u}_\ell} + \sum_j \left(\frac{\partial x}{\partial u_j}, N \right) \frac{\partial^2 u_j}{\partial \tilde{u}_k \partial \tilde{u}_\ell} \\ &= \sum_{i,j} b_{ij} \frac{\partial u_i}{\partial \tilde{u}_k} \frac{\partial u_j}{\partial \tilde{u}_\ell}. \end{aligned}$$

The quadratic form

$$d\sigma^2 := \sum_{1 \leq i,j \leq 2} b_{ij}(N) du_i du_j \left(= \sum_{1 \leq k,\ell \leq 2} \tilde{b}_{k\ell}(N) d\tilde{u}_k d\tilde{u}_\ell \right)$$

is called *the second fundamental form* of M with respect to N. The vector dx^2/ds^2 is called the *curvature vector* of γ, where s is the arclength parameter. As is seen by differentiating the identity $(dx/ds, dx/ds) = 1$ with respect to s, the curvature vector is always perpendicular to the tangent vector dx/ds of γ.

(1.1.2) *For an arbitrarily given regular curve* $\gamma : u_i = u_i(t)$ $(i = 1, 2)$ *in* M, *it holds that*

$$k_\gamma(N) := \left(\frac{d^2x}{ds^2}, N\right) = \frac{d\sigma^2}{ds^2} = \frac{\sum_{i,j} b_{ij} u_i' u_j'}{\sum_{i,j} g_{ij} u_i' u_j'} \qquad (N \in N_{\gamma(t)}(M)).$$

PROOF. Since $(\partial x/\partial u_i, N) = 0$ for $i = 1, 2$, we have

$$\begin{aligned}
\left(\frac{d^2x}{ds^2}, N\right) &= \left(\frac{d}{ds}\left(\sum_i \frac{du_i}{ds}\frac{\partial x}{\partial u_i}\right), N\right) \\
&= \sum_i \frac{d^2u_i}{ds^2}\left(\frac{\partial x}{\partial u_i}, N\right) + \sum_{i,j} \frac{du_i}{ds}\frac{du_j}{ds}\left(\frac{\partial^2 x}{\partial u_i \partial u_j}, N\right) \\
&= \sum_{i,j} b_{ij}(N)\frac{du_i}{ds}\frac{du_j}{ds} \\
&= \left(\frac{dt}{ds}\right)^2 \sum_{i,j} b_{ij}(N)\frac{du_i}{dt}\frac{du_j}{dt}.
\end{aligned}$$

Since $(ds/dt)^2 = \sum_{i,j} g_{i,j} u_i' u_j'$, we have the desired equality.

The assertion (1.1.2) shows that $k_\gamma(N)$ depends only on N and the tangent vector of γ at p. Take a nonzero vector $N \in N_p(M)$ and a unit tangent vector $T \in T_p(M)$. Choose a curve $x(s)$ in M with arclength parameter s such that $x(0) = p$ and $(dx/ds)(0) = T$, and define the *normal curvature* of M in the direction T with respect to the normal vector N by

$$k(N, T) := \left(\frac{d^2x}{ds^2}, N\right).$$

Consider the plane P which includes the vectors N and T and let γ be the curve which is defined as the intersection of P and M. By elementary calculation, we can show that $k(N, T)$ is the reciprocal of radius of curvature for the curve γ in the plane P.

Set

$$\begin{aligned}
&k_1(N) := \max\{k(N, T); T \in T_p(M),\ |T| = 1\}, \\
&k_2(N) := \min\{k(N, T); T \in T_p(M),\ |T| = 1\}.
\end{aligned} \tag{1.1.3}$$

The *mean curvature* of M for the direction N at p is defined by

$$H_p(N) := \frac{k_1(N) + k_2(N)}{2}.$$

The quantities $k_1(N)$ and $k_2(N)$ are the critical values of the second fundamental form $d\sigma^2(T)$ under the condition that $ds^2(T) = 1$. As is easily seen by Lagrange's method of indeterminate coefficients, the quantities $k_1(N)$ and $k_2(N)$ are two roots of the equation

$$\det(b_{ij}(N) - \lambda g_{ij}; 1 \le i, j \le 2) = 0.$$

Hence, we get

$$H_p(N) = \frac{g_{11}b_{22}(N) + g_{22}b_{11}(N) - 2g_{12}b_{12}(N)}{2(g_{11}g_{22} - g_{12}^2)}. \tag{1.1.4}$$

DEFINITION 1.1.5. A surface M is called a *minimal surface* in $\mathbf{R}^m$ if $H_p(N) = 0$ for all $p \in M$ and $N \in N_p(M)$.

We explain here a geometric meaning of minimal surfaces. Let D be a subdomain of M with smooth boundary. For a given normal vector field N on a neighborhood of $\bar{D}$ which vanishes on the boundary ∂D, consider the surfaces

$$D^t : x = x^t := x(u_1, u_2) + tN(u_1, u_2) \qquad ((u_1, u_2) \in D)$$

for a sufficiently small t. Let us denote the first fundamental form of D^t by

$$(ds^t)^2 = g_{11}^t du_1^2 + 2g_{12}^t du_1 du_2 + g_{22}^t du_2^2$$

and the area of D^t by $A(D_t)$. Since

$$\left(\frac{\partial x}{\partial u_i}, \frac{\partial N}{\partial u_j}\right) = \frac{\partial}{\partial u_j}\left(\frac{\partial x}{\partial u_i}, N\right) - \left(\frac{\partial^2 x}{\partial u_j \partial u_i}, N\right) = -b_{ji}(N),$$

we have

$$\begin{aligned} g_{ij}^t &= \left(\frac{\partial x^t}{\partial u_i}, \frac{\partial x^t}{\partial u_j}\right) = \left(\frac{\partial x}{\partial u_i} + t\frac{\partial N}{\partial u_i}, \frac{\partial x}{\partial u_j} + t\frac{\partial N}{\partial u_j}\right) \\ &= g_{ij} - 2tb_{ij}(N) + t^2\left(\frac{\partial N}{\partial u_i}, \frac{\partial N}{\partial u_j}\right), \end{aligned}$$

and so

$$\left.\frac{dg_{ij}^t}{dt}\right|_{t=0} = -2b_{ij}(N).$$

Therefore, we obtain

$$\begin{aligned}
\frac{d}{dt}A(D_t)\Big|_{t=0} &= \frac{d}{dt}\left(\int_D \sqrt{g^t_{11}g^t_{22}-(g^t_{12})^2}du_1du_2\right)\Big|_{t=0} \\
&= \int_D \frac{d}{dt}\sqrt{g^t_{11}g^t_{22}-(g^t_{12})^2}\Big|_{t=0}du_1du_2 \\
&= -\int_D \frac{g_{11}b_{22}(N)+g_{22}b_{11}(N)-2g_{12}b_{12}(N)}{\sqrt{g_{11}g_{22}-g_{12}^2}}du_1du_2.
\end{aligned}$$

By the use of (1.1.4) we conclude that

$$\frac{d}{dt}A(D_t)\Big|_{t=0} = -2\int\int_D H(N)\Omega_{ds^2}, \tag{1.1.6}$$

where $\Omega_{ds^2} := \sqrt{g_{11}g_{22}-g_{12}^2}du_1du_2$ is the area form of M.

As is easily seen by Definition 1.1.5 and (1.1.6), we have

(1.1.7) *A surface M is minimal if and only if*

$$\frac{d}{dt}A(D_t)\Big|_{t=0} = 0$$

for all subdomains D with smooth boundary and normal vector fields N vanishing on ∂D.

According to (1.1.7) we may say that a surface M is minimal if and only if for arbitrary local perturbations of the surface M in the normal direction the area function has a critical value at M. In fact, it is known that the area functional applied to local perturbations of a small piece of a minimal surface M takes a minimal value at M itself(cf., [3]).

We now recall the notion of isothermal coordinates for a surface with a metric ds^2. By definition, a system of local coordinates u_1, u_2 on an open set U in M is called a system of *isothermal coordinates* on U if ds^2 can be represented as

$$ds^2 = \lambda^2(du_1^2 + du_2^2)$$

for some positive C^∞ function λ on U. This means that $g_{11} = g_{22}, g_{12} = 0$ for a representation

$$ds^2 = g_{11}du_1^2 + 2g_{12}du_1du_2 + g_{22}du_2^2.$$

By (1.1.1) local coordinates u, v on a surface $x : M \to \mathbf{R}^m$ are isothermal coordinates if and only if

$$(1.1.8) \qquad \left(\frac{\partial x}{\partial u}, \frac{\partial x}{\partial u}\right) = \left(\frac{\partial x}{\partial v}, \frac{\partial x}{\partial v}\right), \quad \left(\frac{\partial x}{\partial u}, \frac{\partial x}{\partial v}\right) = 0.$$

THEOREM 1.1.9. *For every surface M there is a family of isothermal local coordinates whose domains cover the totality of M.*

The proof is omitted. Refer to e.g. [11].

For a particular case where M is a minimal surface in $\mathbf{R}^m$, we give a simple proof of Theorem 1.1.9 (cf., [63], pp. 31 ~ 32).

Take a point $p_0 \in M$. Changing indices if necessary, we may take the coordinates x_1, x_2 as a system of local coordinates because M is immersed in $\mathbf{R}^m$. Therefore, the surface M is represented as

$$M; \begin{cases} x_i = x_i & i = 1, 2 \\ x_i = f_i(x_1, x_2) & i = 3, \dots, m \end{cases}$$

around p. Consider the vector-valued function $f := (f_3, \dots, f_m)$ and set

$$p := \frac{\partial f}{\partial x_1}, \quad q := \frac{\partial f}{\partial x_2}, \quad r := \frac{\partial^2 f}{\partial x_1^2}, \quad s := \frac{\partial^2 f}{\partial x_1 \partial x_2}, \quad t := \frac{\partial^2 f}{\partial x_2^2}.$$

We then have

$$g_{11} = 1 + |p|^2, \quad g_{12} = g_{21} = (p, q), \quad g_{22} = 1 + |q|^2.$$

Normal vectors $N = (N_1, \dots, N_m)$ of M are represented as

$$N_1 = -\sum_{k=3}^{m} N_k \frac{\partial f_k}{\partial x_1}, \quad N_2 = -\sum_{k=3}^{m} N_k \frac{\partial f_k}{\partial x_2},$$

with arbitrary constants N_i $(3 \leq i \leq m)$ because a vector N is normal if and only if $(N, \partial x/\partial x_1) = (N, \partial x/\partial x_2) = 0$. Since $b_{ij}(N) = \sum_{k=3}^{m} N_k (\partial^2/\partial x_i \partial x_j) f_k$, by the assumption of minimality of M and the formula (1.1.4) it holds that

$$g_{11} \frac{\partial^2 f}{\partial x_2^2} - 2 g_{12} \frac{\partial^2 f}{\partial x_1 \partial x_2} + g_{22} \frac{\partial^2 f}{\partial x_1^2} = 0,$$

namely

$$\Psi := (1+|q|^2)r - 2(p,q)s + (1+|p|^2)t = 0.$$

Set

$$W := \sqrt{g_{11}g_{22} - g_{12}^2} = ((1+|p|^2)(1+|q|^2) - (p,q)^2)^{1/2}.$$

Then, we have

$$(1.1.10) \quad -\frac{\partial}{\partial x_1}\frac{(p,q)}{W} + \frac{\partial}{\partial x_2}\frac{1+|p|^2}{W} = \frac{1}{W^3}\left((p,q)p - (1+|p|^2)q,\ \Psi\right).$$

In fact, as is easily seen by the use of the identities

$$\frac{\partial p}{\partial x_1} = r,\ \frac{\partial p}{\partial x_2} = \frac{\partial q}{\partial x_1} = s,\ \frac{\partial q}{\partial x_2} = t,$$

both sides of (1.1.10) have the same expansions. Similarly, we can prove

$$(1.1.11) \quad \frac{\partial}{\partial x_1}\frac{1+|q|^2}{W} - \frac{\partial}{\partial x_2}\frac{(p,q)}{W} = \frac{1}{W^3}\left((p,q)q - (1+|q|^2)p,\ \Psi\right)$$

Since $\Psi = 0$ by the assumption of minimality of M, (1.1.10) and (1.1.11) show that the forms

$$\omega_1 := \frac{1+|p|^2}{W}dx_1 + \frac{(p,q)}{W}dx_2,\quad \omega_2 := \frac{(p,q)}{W}dx_1 + \frac{1+|q|^2}{W}dx_2,$$

are closed. Therefore, we can choose functions F and G in some neighborhood of p_0 such that $dF = \omega_1$ and $dG = \omega_2$. We now define

$$u := x_1 + F(x_1,x_2), v := x_2 + G(x_1,x_2).$$

Since $J := \dfrac{\partial(u,v)}{\partial(x_1,x_2)} = 2 + \dfrac{|p|^2+|q|^2+2}{W} > 0$, we may take (u,v) as local coordinates in a neighborhood of p_0. Then, for the matrix

$$A := \begin{pmatrix} \frac{\partial x}{\partial u} \\ \frac{\partial x}{\partial v} \end{pmatrix},\quad B := \begin{pmatrix} 1+\frac{1+|p|^2}{W} & \frac{(p,q)}{W} \\ \frac{(p,q)}{W} & 1+\frac{1+|q|^2}{W} \end{pmatrix},$$

we have

$$\begin{aligned} A^t A &= B^{-1}\begin{pmatrix} 1+|p|^2 & (p,q) \\ (p,q) & 1+|q|^2 \end{pmatrix} {}^tB^{-1} \\ &= \frac{W}{J}\begin{pmatrix} 1 & 0 \\ 0 & 1 \end{pmatrix}. \end{aligned}$$

Therefore, we get

$$\left|\frac{\partial x}{\partial u}\right|^2 = \left|\frac{\partial x}{\partial v}\right|^2 = \frac{W}{J}, \quad \left(\frac{\partial x}{\partial u}, \frac{\partial x}{\partial v}\right) = 0.$$

This shows that (u, v) is a system of isothermal coordinates around p_0.

PROPOSITION 1.1.12. *For an oriented surface M with a metric ds^2, if we take two systems of positively oriented isothermal local coordinates (u, v) and (x, y), then $w := u + \sqrt{-1}v$ is a holomorphic function in $z := x + \sqrt{-1}y$ on the common domain of definition.*

PROOF. By assumption, there exists a positive differentiable function λ such that

$$du^2 + dv^2 = \lambda^2(dx^2 + dy^2).$$

Therefore, we have

$$A := u_x^2 + v_x^2 = u_y^2 + v_y^2, \quad u_x u_y + v_x v_y = 0.$$

This means that the Jacobi matrix $\mathbf{J} := \begin{pmatrix} u_x & v_x \\ u_y & v_y \end{pmatrix}$ satisfies the identity $\mathbf{J}\,{}^t\mathbf{J} = A\,\mathbf{I}_2$, where $\mathbf{I}_2$ denotes the unit matrix of degree 2. We then have

$$\mathbf{J}^{-1} = \frac{1}{\det \mathbf{J}} \begin{pmatrix} v_y & -v_x \\ -u_y & u_x \end{pmatrix} = \frac{1}{A} \begin{pmatrix} u_x & u_y \\ v_x & v_y \end{pmatrix}.$$

On the other hand, since $(\det \mathbf{J})^2 = \det(\mathbf{J}\,{}^t\mathbf{J}) = \det(A\mathbf{I}_2) = A^2$ and $\det \mathbf{J} > 0$, we have $A = \det \mathbf{J}$. These imply that $u_x = v_y$ and $u_y = -v_x$. Therefore, the function $w = u + \sqrt{-1}v$ is holomorphic in z.

Let $x : M \to \mathbf{R}^m$ be an oriented surface with a Riemannian metric ds^2. With each positive isothermal local coordinate (u, v) we associate the complex function $z := u + \sqrt{-1}v$. By virtue of Proposition 1.1.12, the surface M has a complex structure so that these complex-valued functions define holomorphic local coordinates on M, and so M may be considered as a Riemann surface. Then the metric ds^2 is given by

$$ds^2 = \lambda_z^2(du^2 + dv^2),$$

where

$$\lambda_z^2 := \left(\frac{\partial x}{\partial u}, \frac{\partial x}{\partial u}\right) = \left(\frac{\partial x}{\partial v}, \frac{\partial x}{\partial v}\right).$$

By the use of complex differentiations

$$\frac{\partial x_i}{\partial z} := \frac{1}{2}\left(\frac{\partial x_i}{\partial u} - \sqrt{-1}\frac{\partial x_i}{\partial v}\right), \frac{\partial x_i}{\partial \bar{z}} := \overline{\frac{\partial x_i}{\partial z}}$$

we may rewrite the metric as

$$ds^2 = 2\left(\left|\frac{\partial x_1}{\partial z}\right|^2 + \cdots + \left|\frac{\partial x_m}{\partial z}\right|^2\right)|dz|^2. \tag{1.1.13}$$

We denote by Δ, or more precisely by Δ_z the Laplacian $\dfrac{\partial^2}{\partial u^2} + \dfrac{\partial^2}{\partial v^2}$ in terms of the holomorphic coordinate $z = u + \sqrt{-1}v$. If we take another holomorphic local coordinate ζ, then we have $\Delta_\zeta = |dz/d\zeta|^2\Delta_z$. Since $\lambda_\zeta = \lambda_z|dz/d\zeta|$, the operator $\mathbf{\Delta} := (1/\lambda_z^2)\Delta_z$ does not depend on the choice of the holomorphic local coordinate z, which is called the Laplace-Bertrami operator.

PROPOSITION 1.1.14. *It holds that*
(i) $(\mathbf{\Delta}\, x, X) = 0$ *for each* $X \in T_p(M)$,
(ii) $(\mathbf{\Delta}\, x, N) = 2H(N)$ *for each* $N \in N_p(M)$.

PROOF. By the assumptions, we have

$$\lambda^2 := \left(\frac{\partial x}{\partial u}, \frac{\partial x}{\partial u}\right) = \left(\frac{\partial x}{\partial v}, \frac{\partial x}{\partial v}\right), \left(\frac{\partial x}{\partial u}, \frac{\partial x}{\partial v}\right) = 0.$$

Differentiating these identities, we have

$$\left(\frac{\partial^2 x}{\partial u^2}, \frac{\partial x}{\partial u}\right) = \left(\frac{\partial^2 x}{\partial u \partial v}, \frac{\partial x}{\partial v}\right), \left(\frac{\partial^2 x}{\partial v \partial u}, \frac{\partial x}{\partial v}\right) + \left(\frac{\partial x}{\partial u}, \frac{\partial^2 x}{\partial v^2}\right) = 0.$$

These imply

$$\left(\Delta x, \frac{\partial x}{\partial u}\right) = \left(\frac{\partial^2 x}{\partial u \partial v}, \frac{\partial x}{\partial v}\right) - \left(\frac{\partial^2 x}{\partial v \partial u}, \frac{\partial x}{\partial v}\right) = 0.$$

By a similar method, we have

$$\left(\Delta x, \frac{\partial x}{\partial v}\right) = 0.$$

Since $\partial x/\partial u$ and $\partial x/\partial v$ generate the tangent plane, we conclude the assertion (i) of Proposition 1.1.14. On the other hand, for every normal vector N to M it holds that

$$H(N) = \frac{b_{11}(N) + b_{22}(N)}{2\lambda^2} = \frac{(\varDelta\, x, N)}{2}$$

because $g_{11} = g_{22} = \lambda^2$ and $g_{12} = 0$, which shows (ii) of Proposition 1.1.14.

THEOREM 1.1.15. *Let $x = (x_1, \dots, x_m) : M \to \mathbf{R}^m$ be a surface immersed in $\mathbf{R}^m$, which is considered as a Riemann surface as above. Then, M is minimal if and only if each x_i is a harmonic function on M, namely,*

$$\varDelta_z x_i \equiv \left(\frac{\partial^2}{\partial u^2} + \frac{\partial^2}{\partial v^2}\right) x_i = 0 \qquad (1 \le i \le m)$$

for every holomorphic local coordinate $z = u + \sqrt{-1}v$.

PROOF. By Proposition 1.1.14, (i), $\varDelta x = 0$ if and only if $\varDelta x$ is perpendicular to the normal space of M. This is equivalent to the condition $H = 0$ by Proposition 1.1.14, (ii).

COROLLARY 1.1.16. *There exists no compact minimal surface without boundary in $\mathbf{R}^m$.*

PROOF. For a minimal surface $x = (x_1, \dots, x_m) : M \to \mathbf{R}^m$ immersed in $\mathbf{R}^m$, if M is compact, then each x_i takes the maximum values at a point in M. By the maximum principle of harmonic functions, x_i is a constant. This is impossible because x is an immersion.

§1.2 The Gauss map of minimal surfaces in $\mathbf{R}^m$

We next explain the Gauss map of a surface $x = (x_1, x_2, \dots, x_m) : M \to \mathbf{R}^m$ immersed in $\mathbf{R}^m$.

Firstly, consider the set of all oriented 2-planes in $\mathbf{R}^m$ which contain the origin and denote it by Π. To clarify the set Π, we regard it as a subset of the $(m-1)$-dimensional complex projective space $P^{m-1}(\mathbf{C})$ as follows. To each $P \in \Pi$, taking a positively oriented basis $\{X, Y\}$ of P such that

$$|X| = |Y|, \quad (X,\ Y) = 0, \qquad (1.2.1)$$

we assign the point $\Phi(P) = \pi(X - \sqrt{-1}Y)$, where π denotes the canonical projection of $\mathbf{C}^m - \{0\}$ onto $P^{m-1}(\mathbf{C})$, namely, the map which maps each $p = (w_1, \dots, w_m) \neq (0, \dots, 0)$ to the equivalence class

$$(w_1 : \cdots : w_m) := \{(cw_1, \dots, cw_m); c \in \mathbf{C} - \{0\}\}.$$

For another positive basis $\{\tilde{X}, \tilde{Y}\}$ of P satisfying (1.2.1) we can find a real number θ such that

$$\begin{aligned} \tilde{X} &= r(\cos\ \theta\ X + \sin\ \theta\ Y), \\ \tilde{Y} &= r(-\sin\ \theta\ X + \cos\ \theta\ Y), \end{aligned}$$

where $r := |\tilde{X}|/|X|$. Therefore, we can write

$$\tilde{X} - \sqrt{-1}\tilde{Y} = re^{i\theta}(X - \sqrt{-1}Y).$$

This shows that the value $\Phi(P)$ does not depend on the choice of a positive basis of P satisfying (1.2.1) but only on P. On the other hand, $\Phi(P)$ is contained in the quadric

$$Q_{m-2}(\mathbf{C}) := \{(w_1 : \cdots : w_m); w_1^2 + \cdots + w_m^2 = 0\}(\subset P^{m-1}(\mathbf{C})).$$

In fact, for a positive basis $\{X, Y\}$ satisfying (1.2.1) we have

$$(X - \sqrt{-1}Y,\ X - \sqrt{-1}Y) = (X, X) - 2\sqrt{-1}(X, Y) - (Y, Y) = 0.$$

Conversely, take an arbitrary point $Q \in Q_{m-2}(\mathbf{C})$. If we choose some $W \in \mathbf{C}^m - \{0\}$ with $\pi(W) = Q$ and write $W = X - \sqrt{-1}Y$ with real vectors X and Y, then X and Y satisfy the condition (1.2.1) and Φ maps the oriented 2-plane P with positive basis $\{X, Y\}$ to the point Q. This shows that Φ is surjective. Moreover, it is not difficult to show that Φ is injective. In the following sections, there will be no confusion if we identity Π with $Q_{m-2}(\mathbf{C})$.

Now, consider a surface $x = (x_1, x_2, \cdots, x_m) : M \to \mathbf{R}^m$ immersed in $\mathbf{R}^m$. For each point $p \in M$, the oriented tangent plane $T_p(M)$ is canonically identified with an element of Π after the parallel translation which maps p to the origin.

DEFINITION 1.2.2. The *Gauss map* of a surface M is defined as the map of M into $Q_{m-2}(\mathbf{C})$ which maps each point $p \in M$ to $\Phi(T_p(M))$.

Usually, the Gauss map is defined as the conjugate of the Gauss map defined as above. We adopt Definition 1.2.2 for convenience' sake for simplifying the description of function-theoretic properties of minimal surfaces.

For a system of positively oriented isothermal local coordinates (u, v) the vectors

$$X = \frac{\partial x}{\partial u}, \quad Y = \frac{\partial x}{\partial v}$$

give a positive basis of $T_p(M)$ satisfying the condition (1.2.1) because of (1.1.8). Therefore, the Gauss map of M is locally given by

$$G = \pi(X - \sqrt{-1}Y) = \left(\frac{\partial x_1}{\partial z} : \frac{\partial x_2}{\partial z} : \cdots : \frac{\partial x_m}{\partial z}\right), \tag{1.2.3}$$

where $z = u + \sqrt{-1}v$. We may write $G = (\omega_1 : \cdots : \omega_m)$ with globally defined holomorphic forms $\omega_i := \partial x_i \equiv (\partial x_i/\partial z)dz$.

We have the following criterion for minimal surfaces.

PROPOSITION 1.2.4 (cf., [46, Theorem 1.1]). *A surface $x : M \to \mathbf{R}^m$ is minimal if and only if the Gauss map $G : M \to P^{m-1}(\mathbf{C})$ is holomorphic.*

PROOF. Assume that M is minimal. We then have

$$\frac{\partial}{\partial \bar{z}}\left(\frac{\partial x}{\partial z}\right) = \frac{1}{4}\Delta x = 0$$

by Theorem 1.1.15. This shows that $\partial x/\partial z$ satisfies Cauchy-Riemann's equation. Hence, the Gauss map G is holomorphic.

Conversely, assume that G is holomorphic. The problem is local. For a holomorphic local coordinate z we set $f_i := \dfrac{\partial x_i}{\partial z}$ $(1 \leq i \leq m)$. After a suitable change of indices, we may assume that f_m has no zero. Since f_i/f_m are holomorphic, we have

$$\begin{aligned}\frac{1}{4}\Delta x_i &= \frac{\partial^2 x_i}{\partial z \partial \bar{z}} = \frac{\partial}{\partial \bar{z}}\left(\frac{f_i}{f_m} f_m\right) \\ &= \frac{\partial}{\partial \bar{z}}\left(\frac{f_i}{f_m}\right) f_m + \frac{f_i}{f_m}\frac{\partial f_m}{\partial \bar{z}} = f_i \frac{1}{f_m}\frac{\partial f_m}{\partial \bar{z}}\end{aligned}$$

for $i = 1, 2, \cdots, m$. Write

$$\frac{1}{f_m}\frac{\partial f_m}{\partial \bar{z}} = h_1 + \sqrt{-1}h_2$$

with real-valued functions h_1, h_2 and take the real parts of both sides of the above equation to see

$$\Delta x = 2\left(\frac{\partial x}{\partial u} h_1 + \frac{\partial x}{\partial v} h_2\right) \in T_p(M).$$

According to Proposition 1.1.14, (i), we obtain $(\Delta x, \Delta x) = 0$ and so $\Delta x = 0$. This implies that M is a minimal surface by virtue of Theorem 1.1.15.

We say that a holomorphic form ω on a Riemann surface M has no real period if

$$\operatorname{Re} \int_\gamma \omega = 0$$

for every closed curve γ in M. If ω has no real period, then the quantity

$$x(z) = \operatorname{Re} \int_{\gamma_{z_0}^z} \omega$$

depends only on z and z_0 for a piecewise smooth curve $\gamma_{z_0}^z$ in M joining z_0 and z and so x is a well-defined function of z on M, which we denote by

$$x(z) = \operatorname{Re} \int_{z_0}^z \omega$$

in the following. Related to Proposition 1.2.4, we state here the following construction theorem of minimal surfaces.

THEOREM 1.2.5. *Let M be an open Riemann surface and let $\omega_1, \omega_2, \ldots, \omega_m$ be holomorphic forms on M such that they have no common zero, no real periods and locally satisfy the identity*

$$f_1^2 + f_2^2 + \cdots + f_m^2 = 0 \tag{1.2.6}$$

for holomorphic functions f_i with $\omega_i = f_i dz$. Set

$$x_i = 2 \operatorname{Re} \int_{z_0}^z \omega_i, \tag{1.2.7}$$

for an arbitrarily fixed point z_0 of M. Then, the surface $x = (x_1, \ldots, x_m) : M \to \mathbf{R}^m$ is a minimal surface immersed in $\mathbf{R}^m$ such that the Gauss map *is the map* $G = (\omega_1 : \cdots : \omega_m) : M \to Q_{m-2}(\mathbf{C})$ *and the induced metric is given by*

$$ds^2 = 2(|\omega_1|^2 + \cdots + |\omega_m|^2). \tag{1.2.8}$$

PROOF. By assumption, the x_i are well-defined single-valued functions on M. Consider the map $x := (x_1, \dots, x_m) : M \to \mathbf{R}^m$. Since $\partial x_i/\partial z = f_i$ $(1 \le i \le m)$, by (1.2.6) we have

$$4\sum_{i=1}^{m} f_i^2 = \left(\frac{\partial x}{\partial u}, \frac{\partial x}{\partial u}\right) - 2\sqrt{-1}\left(\frac{\partial x}{\partial u}, \frac{\partial x}{\partial v}\right) - \left(\frac{\partial x}{\partial v}, \frac{\partial x}{\partial v}\right) = 0$$

for $z = u + \sqrt{-1}v$. This gives that

$$\left(\frac{\partial x}{\partial u}, \frac{\partial x}{\partial u}\right) = \left(\frac{\partial x}{\partial v}, \frac{\partial x}{\partial v}\right), \quad \left(\frac{\partial x}{\partial u}, \frac{\partial x}{\partial v}\right) = 0.$$

Moreover,

$$\begin{aligned}\sum_{i<j}\left|\frac{\partial(x_i, x_j)}{\partial(u, v)}\right|^2 &= \left(\frac{\partial x}{\partial u}, \frac{\partial x}{\partial u}\right)\left(\frac{\partial x}{\partial v}, \frac{\partial x}{\partial v}\right) - \left(\frac{\partial x}{\partial u}, \frac{\partial x}{\partial v}\right)^2 \\ &= 4(|f_1|^2 + \cdots + |f_m|^2)^2 > 0.\end{aligned}$$

Hence, x is an immersion. Then, the induced metric is given by

$$ds^2 = \left(\frac{\partial x}{\partial u}, \frac{\partial x}{\partial u}\right)(du^2 + dv^2) = 2(|f_1|^2 + \cdots + |f_m|^2)|dz|^2.$$

and (u, v) gives a system of isothermal coordinates for the induced metric ds^2. On the other hand, by (1.2.3) the Gauss map G of M is given by $G = (f_1 : \cdots : f_m)$ with holomorphic functions f_i and so holomorphic. According to Proposition 1.2.4, the surface M is a minimal surface. Theorem 1.2.5 is completely proved.

Let M be a Riemann surface with a metric ds^2 which is conformal, namely, represented as

$$ds^2 = \lambda_z^2|dz|^2 \tag{1.2.9}$$

with a positive C^∞ function λ_z in term of a holomorphic local coordinate z.

DEFINITION 1.2.10. For each point $p \in M$ we define the *Gaussian curvature* of M at p by

$$K \equiv K_{ds^2} := \mathit{\Delta}_z \log \lambda_z \left(= \frac{\Delta_z \log \lambda_z}{\lambda_z^2} \right) \tag{1.2.11}$$

For a minimal surface M immersed in $\mathbf{R}^m$ consider the system of holomorphic functions $\tilde{G} = (f_1, \ldots, f_m)$ for $f_i := \partial x_i / \partial z$ and set $|\tilde{G}| := (|f_1|^2 + \cdots + |f_m|^2)^{1/2}$. In view of (1.2.8), we have easily

$$K_{ds^2} = \frac{2}{|\tilde{G}|^2} \frac{\partial^2}{\partial \bar{z} \partial z} \log |\tilde{G}| = \frac{1}{|\tilde{G}|^6} \left(\sum_{i<j} |f_i f_j' - f_j f_i'|^2 \right). \tag{1.2.12}$$

This implies that the curvature of a minimal surface is always nonpositive.

By definition, a surface with a metric is *flat* if and only if the Gaussian curvature vanishes identically. By (1.2.12) this means that $f_i / f_{i_0} \equiv$ const. $(1 \leq i \leq m)$ for some i_0 with $f'_{i_0} \not\equiv 0$ and, therefore, that the Gauss map G is a constant.

PROPOSITION 1.2.13. *For a minimal surface M immersed in $\mathbf{R}^m$, M is flat, or equivalently, the Gauss map of M is a constant if and only if it lies in a plane.*

PROOF. The Gauss map of a surface which lies in a plane is obviously a constant. Conversely, we assume that the Gauss map $G := (g_1 : \cdots : g_m)$ is a constant. This means that every tangent plane $T_p(M)$ of M $(p \in M)$ is perpendicular to $(m-2)$ particular linearly independent normal vectors $N_1, \ldots, N_{m-2}$. We then have

$$\left(\frac{\partial x}{\partial u}, N_k \right) = \left(\frac{\partial x}{\partial v}, N_k \right) = 0 \quad (1 \leq i \leq m-2)$$

for all local coordinates u and v. Therefore, each (x, N_k) is a constant for $k = 1, \ldots, m-2$ and so M lies in a plane.

§1.3 Enneper-Weierstrass representations of minimal surfaces in $\mathbf{R}^3$

We shall study the Gauss map of minimal surfaces in $\mathbf{R}^3$ more concretely. In $\mathbf{R}^3$, each oriented plane $P \in \Pi$ is uniquely determined by the unit vector N such that it is perpendicular to P and the system $\{X, Y, N\}$ is a positive orthonormal basis of $\mathbf{R}^3$ for arbitrarily chosen positively oriented orthonormal basis $\{X, Y\}$ of P. For an oriented surface in $\mathbf{R}^3$ the tangent plane is uniquely determined by the positively oriented unit normal vector. On the other hand, the unit sphere S^2 of all unit vectors in $\mathbf{R}^3$ is bijectively mapped onto the extended complex plane $\bar{\mathbf{C}} = \mathbf{C} \cup \{\infty\}$ by the stereographic projection ϖ.

We give here the following:

DEFINITION 1.3.1. For a minimal surface M immersed in $\mathbf{R}^3$ the *classical Gauss map* $g : M \to \bar{\mathbf{C}}$ of M is defined as the map which maps each point $p \in M$ to the point $\varpi(N_p) \in \bar{\mathbf{C}}$, where N_p is the positively oriented unit normal vector N_p of M at p.

We shall study more precisely the relation between the Gauss map and the classical Gauss map of a surface in $\mathbf{R}^3$.

Let us begin by studying the stereographic projection ϖ. For an arbitrary point $n = (\xi, \eta, \zeta) \in S^2$ set $z = x + \sqrt{-1}y := \varpi(n)$, which means that, if $n \neq (0,0,1)$, then the points $(0,0,1)$, (ξ, η, ζ) and $(x, y, 0)$ are collinear and, otherwise, $z = \infty$. Then, by elementary calculations, we see

$$\xi = \frac{z + \bar{z}}{|z|^2 + 1}, \quad \eta = \sqrt{-1}\frac{\bar{z} - z}{|z|^2 + 1}, \quad \zeta = \frac{|z|^2 - 1}{|z|^2 + 1}. \tag{1.3.2}$$

For two points $n_1 = (\xi_1, \eta_1, \zeta_1)$ and $n_2 = (\xi_2, \eta_2, \zeta_2) \in S^2$ we denote by θ $(0 \leq \theta \leq \pi)$ the angle between vectors n_1 and n_2 and set $\alpha := \varpi(n_1)$, $\beta := \varpi(n_2)$. Define

$$|\alpha, \beta| = \sin \frac{\theta}{2}. \tag{1.3.3}$$

Geometrically, $2|\alpha, \beta|$ is the chordal distance between α and β. As is easily shown, if $\alpha \neq \infty$ and $\beta \neq \infty$, then

$$|\alpha, \beta| = \frac{|\alpha - \beta|}{\sqrt{1 + |\alpha|^2}\sqrt{1 + |\beta|^2}}$$

and, if either α or β, say β, is equal to ∞, then $|\alpha, \beta| = 1/\sqrt{1 + |\alpha|^2}$.

Now, we take an arbitrary point $(w_1 : w_2 : w_3) \in Q_1(\mathbf{C})$. Write $w_i = x_i - \sqrt{-1}y_i$ $(1 \leq i \leq 3)$ with real numbers x_i, y_i and set

$$W := (w_1, w_2, w_3), \ X := (x_1, x_2, x_3), \ Y := (y_1, y_2, y_3).$$

Since $w_1^2 + w_2^2 + w_3^2 = 0$, they satisfy the condition (1.2.1). Multiplying W by some nonzero constant, we may assume that $|X| = |Y| = 1$. Then, the unit normal vector of the plane which has a positive basis $\{X, Y\}$ is given by

$$N := X \times Y = \mathrm{Im}\{w_2\bar{w}_3,\ w_3\bar{w}_1,\ w_1\bar{w}_2\}.$$

For the case where $w_1 \neq \sqrt{-1}w_2$, we assign to W the point

$$z := \frac{w_3}{w_1 - \sqrt{-1}w_2} \tag{1.3.4}$$

and, otherwise, the point $z := \infty$. This correspondence is continuous inclusively at ∞. To see this, rewrite (1.3.4) as

$$z = \frac{w_3(w_1 + \sqrt{-1}w_2)}{w_1^2 + w_2^2} = -\frac{w_1 + \sqrt{-1}w_2}{w_3}$$

and observe that z tends to ∞ if W tends to the point with $w_1 = \sqrt{-1}w_2$ because $w_3 = 0$ and $w_1 + \sqrt{-1}w_2 \neq 0$ for this point. Since

$$\frac{1}{z} = \frac{w_1 - \sqrt{-1}w_2}{w_3},$$

we have

$$\frac{w_1}{w_3} = \frac{1}{2}\left(\frac{1}{z} - z\right), \quad \frac{w_2}{w_3} = \frac{\sqrt{-1}}{2}\left(\frac{1}{z} + z\right).$$

Since $|w_1|^2 + |w_2|^2 + |w_3|^2 = |X|^2 + |Y|^2 = 2$, one has

$$|w_3|^2 = \frac{2}{|w_1/w_3|^2 + |w_2/w_3|^2 + 1} = \frac{4|z|^2}{(|z|^2+1)^2}.$$

These yield that

$$\begin{aligned} N &= |w_3|^2\ \mathrm{Im}\left(\frac{w_2}{w_3}, \overline{\left(\frac{w_1}{w_3}\right)}, \frac{w_1}{w_3}\overline{\left(\frac{w_2}{w_3}\right)}\right) \\ &= \left(\frac{2\ \mathrm{Re}\ z}{|z|^2+1}, \frac{2\ \mathrm{Im}\ z}{|z|^2+1}, \frac{|z|^2-1}{|z|^2+1}\right). \end{aligned}$$

By (1.3.2) this shows that the point in S^2 corresponding to $W \in Q_1(\mathbf{C})$ is mapped to the the above point z by the stereographic projection.

Now, we go back to the study of surfaces in $\mathbf{R}^3$. Let $x = (x_1, x_2, x_3) : M \to \mathbf{R}^3$ be a surface immersed in $\mathbf{R}^3$ whose Gauss map G is not a constant. As is stated in the previous section, M may be considered as a Riemann surface with a conformal metric ds^2. For a holomorphic local coordinate $z = u + \sqrt{-1}v$, G is represented as $G = (\omega_1 : \omega_2 : \omega_3) = (f_1 : f_2 : f_3)$, where

$$\omega_i = f_i dz = \partial x_i \qquad (1 \leq i \leq 3). \tag{1.3.5}$$

Set

$$\omega := \omega_1 - \sqrt{-1}\omega_2 (\not\equiv 0), \; g := \frac{f_3}{f_1 - \sqrt{-1}f_2}. \tag{1.3.6}$$

As is easily seen by the above discussion, the function g is just the classical Gauss map of M.

Since the above correspondences are all biholomorphic, we see easily by Proposition 1.2.4 the following:

PROPOSITION 1.3.7. *For a surface M immersed in $\mathbf{R}^3$, M is a minimal surface if and only if the classical Gauss map is meromorphic on M.*

We explain here the following Enneper-Weierstrass representation theorem for minimal surfaces.

THEOREM 1.3.8. *Let $x = (x_1, x_2, x_3) : M \to \mathbf{R}^3$ be a nonflat minimal surface immersed in $\mathbf{R}^3$. Consider the holomorphic forms $\omega_1, \omega_2, \omega_3, \omega$ and the meromorphic function g which is defined by (1.3.5) and (1.3.6) respectively. Then,*

(i) it holds that

$$\omega_1 = \frac{1}{2}(1 - g^2)\omega, \; \omega_2 = \frac{\sqrt{-1}}{2}(1 + g^2)\omega, \; \omega_3 = g\omega \tag{1.3.9}$$

and

$$\begin{aligned} x_1 &= \mathrm{Re} \int_{z_0}^{z} (1 - g^2)\omega + \text{const.} \\ x_2 &= \mathrm{Re}\sqrt{-1} \int_{z_0}^{z} (1 + g^2)\omega + \text{const.} \\ x_3 &= \mathrm{Re}\; 2 \int_{z_0}^{z} g\omega + \text{const.} \end{aligned} \tag{1.3.10}$$

(ii) *the metric induced from the standard metric on* $\mathbf{R}^3$ *is given by*

$$ds^2 = (1 + |g|^2)^2 |\omega|^2. \tag{1.3.11}$$

(iii) *the holomorphic form* ω *has a zero of order* $2k$ *when and only when* g *has a pole of order* k.

PROOF. Consider the functions f_i and h with $\omega_i = f_i dz$ and $\omega = h dz$ for a holomorphic local coordinate z. Obviously, $gh = f_3$. Since $f_1^2 + f_2^2 + f_3^2 = 0$, we have

$$\begin{aligned}
\frac{1}{2}(1 - g^2)h &= \frac{1}{2}\left(1 - \left(\frac{f_3}{f_1 - \sqrt{-1}f_2}\right)^2\right)(f_1 - \sqrt{-1}f_2) \\
&= \frac{f_1^2 - f_2^2 - 2\sqrt{-1}f_1 f_2 - f_3^2}{2(f_1 - \sqrt{-1}f_2)} \\
&= f_1.
\end{aligned}$$

Similarly,

$$\frac{\sqrt{-1}}{2}(1 + g^2)h = \frac{\sqrt{-1}}{2}\frac{f_1^2 - f_2^2 - 2\sqrt{-1}f_1 f_2 + f_3^2}{f_1 - \sqrt{-1}f_2} = f_2,$$

On the other hand, for $i = 1, 2, 3$,

$$2 \operatorname{Re} \int_{z_0}^{z} f_i(\zeta)d\zeta = \int_{z_0}^{z} \frac{\partial x_i}{\partial u}du + \frac{\partial x_i}{\partial v}dv = x_i(z) - x_i(z_0),$$

which shows the representation (1.3.10).

The assertion (ii) is shown by the direct calculations

$$\begin{aligned}
ds^2 &= 2(|f_1|^2 + |f_3|^2 + |f_3|^2)|dz|^2 \\
&= \frac{1}{2}(|1 - g^2|^2 + |1 + g^2|^2 + 4|g^2|^2)|h|^2|dz|^2 \\
&= (1 + |g|^2)^2|h|^2|dz|^2.
\end{aligned}$$

If h has a zero at a point p where g is holomorphic, then f_1, f_2, f_3 have a common zero at p. On the other hand, if g has a pole of order k at a point p, then h has a zero of exact order $2k$ at p, because otherwise some f_i has a pole or f_i's have a common zero. Thus, the assertion (iii) holds.

We can show the following restatement of Theorem 1.2.5 for a particular case $m = 3$.

THEOREM 1.3.12. *Let M be an open Riemann surface, ω a nonzero holomorphic form and g a nonconstant meromorphic function on M. Assume that ω has a zero of order $2k$ when and only when g has a pole of order k and that the holomorphic forms $\omega_1, \omega_2, \omega_3$ defined by (1.3.9) have no real periods. Then, for the functions x_1, x_2, x_3 defined by (1.3.10), the surface*

$$x = (x_1, x_2, x_3) : M \to \mathbf{R}^3$$

is a minimal surface immersed in $\mathbf{R}^3$ whose classical Gauss map is the map g and whose induced metric is given by (1.3.11).

PROOF. This is obvious because the holomorphic forms $\omega_1, \omega_2, \omega_3$ satisfy all assumptions of Theorem 1.2.5.

As is easily seen from (1.2.11) and the assertion (ii) of Theorem 1.3.8, the Gaussian curvature of M is given by

$$K_{ds^2}(p) = -\frac{4|g'|^2}{|h|^2(1+|g|^2)^4}, \tag{1.3.13}$$

where $\omega = hdz$.

The Gaussian curvature of a surface in $\mathbf{R}^3$ at a point p is classically defined as the product of the quantities $k_1(N)$ and $k_2(N)$ given by (1.1.3) for the uniquely determined normal vector N of M at p compatible with the orientation. It is known that the Gaussian curvature in this sense coincides with the curvature defined as in Definition 1.2.11, namely,

$$K(p) = k_1(N)k_2(N) \tag{1.3.14}$$

holds.

We shall give here a proof of (1.3.14) for a particular case where the surface is minimal(cf., [63, Lemma 9.1]). To this end, take a curve

$$\gamma : \zeta = \zeta(s) = u(s) + \sqrt{-1}v(s)$$

with arclength parameter s, where (u, v) is a system of positive isothermal coordinates. By using the identity $\Delta x = 0$, we see easily

$$\begin{aligned} &\mathrm{Re}\left(\frac{\partial^2 x}{\partial z^2}(u' + \sqrt{-1}v')^2\right) \\ &\quad = \frac{1}{2}\left(\frac{\partial^2 x}{\partial u^2}u'^2 + 2\frac{\partial^2 x}{\partial u \partial v}u'v' + \frac{\partial^2 x}{\partial v^2}v^2\right). \end{aligned}$$

Therefore, for the coefficients b_{ij} of the second fundamental form it holds that

$$\sum_{i,j} b_{ij}(N)u_i'u_j' = \left(2\,\mathrm{Re}\left(\frac{\partial^2 x}{\partial z^2}\left(\frac{d\zeta}{ds}\right)^2\right), N\right),$$

where N denotes the normal vector of M. On the other hand, if we consider the functions h and g satisfying the condition (1.3.10) for $\omega = hdz$ we have

$$\begin{aligned}
\frac{\partial^2 x_1}{\partial z^2} &= \frac{1}{2}((1-g^2)h' - 2gg'h) \\
\frac{\partial^2 x_2}{\partial z^2} &= \frac{\sqrt{-1}}{2}((1+g^2)h' + 2gg'h) \\
\frac{\partial^2 x_3}{\partial z^2} &= gh' + g'h
\end{aligned}$$

and the normal vector is given by

$$N = \left(\frac{2\ \mathrm{Re}\ g}{|g|^2+1}, \frac{2\ \mathrm{Im}\ g}{|g|^2+1}, \frac{|g|^2-1}{|g|^2+1}\right).$$

Therefore, by direct calculations we obtain

$$\left(2\frac{\partial^2 x}{\partial z^2}, N\right) = -2g'h.$$

Concludingly, we obtain

$$k_\gamma(N) = -\frac{1}{\lambda^2}\sum_{ij} b_{ij}(N)u_i'u_j' = -\frac{1}{\lambda^2}\mathrm{Re}\left(-2hg'\left(\frac{d\zeta}{ds}\right)^2\right),$$

where $\lambda^2 = |h|^2(1+|g|^2)^2$. On the other hand, $d\zeta/ds$ takes all numbers whose absolute values are one. The maximum and minimum of the possible values of the $\lambda^2 k_\gamma(N)$'s are $2|hg'|^2$ and $-2|hg'|^2$. Therefore, we can conclude the equality

$$k_1(N)k_2(N) = -\frac{4|g'|^2}{|h|^2(1+|g|^2)^4},$$

which is just the desired identity (1.3.14).

§1.4 Sum to product estimates for meromorphic functions

In this section, we shall give some properties of meromorphic functions on a disc in $\mathbf{C}$ for later use.

Take q (≥ 2) mutually distinct numbers $\alpha_1, \dots, \alpha_q$ in $\bar{\mathbf{C}} = \mathbf{C} \cup \{\infty\}$. Set

$$L := \min_{1 \leq i < j \leq q} |\alpha_i, \alpha_j|. \tag{1.4.1}$$

Then, we have the following:

(1.4.2). *For all $w \in \bar{\mathbf{C}}$ it holds that*

$$|w, \alpha_i| \geq \frac{L}{2}$$

for all except at most one α_i.

In fact, if there exists some $w \in \bar{\mathbf{C}}$ such that $|w, \alpha_i| < L/2$ holds for two distinct indices $i = i_1, i_2$, then we have an absurd conclusion

$$L \leq |\alpha_{i_1}, \alpha_{i_2}| \leq |\alpha_{i_1}, w| + |w, \alpha_{i_2}| < L.$$

Let g be a nonconstant meromorphic function on $\Delta_R := \{z \in \mathbf{C}; |z| < R\}$ $(0 < R \leq +\infty)$ and let $\eta_1, \dots, \eta_q$ be real numbers such that $0 < \eta_j \leq 1$ $(1 \leq j \leq q)$ and

$$\gamma := \eta_1 + \cdots + \eta_q > 1.$$

We can prove the following:

Proposition 1.4.3. *For each ρ with $\rho > 0$ and η with $\gamma - 1 > \gamma\eta \geq 0$, choose a constant $a_0 (\geq e^2)$ satisfying the condition*

$$\frac{1}{\log^2 a_0} + \frac{1}{\log a_0} < \rho' \tag{1.4.4}$$

for $\rho' := \rho/\gamma$. Then, it holds that

$$\Delta \log \frac{(1 + |g|^2)^\rho}{\prod_{j=1}^q \log^{\eta_j} \dfrac{a_0}{|g, \alpha_j|^2}}$$

$$\geq C_1^2 \frac{|g'|^2}{(1 + |g|^2)^2} \prod_{j=1}^q \left(\frac{1}{|g, \alpha_j|^2 \log^2 \dfrac{a_0}{|g, \alpha_j|^2}} \right)^{\eta_j(1-\eta)},$$

where

$$C_1 := 2\left(\frac{L}{2}\log\frac{4a_0}{L^2}\right)^{\gamma-1-\gamma\eta}. \tag{1.4.5}$$

For the proof, we need some lemmas. We first state the following elementary inequality without proof (cf., e.g., [4, p. 7]).

(1.4.6) *For every positive numbers* $x_1, \dots, x_n$ and $a_1, \dots, a_n$,

$$\frac{a_1x_1+\cdots+a_nx_n}{a_1+\cdots+a_n} \geq (x_1^{a_1}\dots x_n^{a_n})^{\frac{1}{a_1+\cdots+a_n}}.$$

LEMMA 1.4.7. *Take nonnegative numbers* $A_1, \dots, A_q$ *and a positive constant* M *such that* $M \geq A_j$ *for all* j *except at most one. Then, for every* η *with* $\gamma - 1 > \gamma\eta \geq 0$,

$$\eta_1A_1 + \eta_2A_2 + \cdots + \eta_qA_q \geq \frac{1}{M^{\gamma-1-\gamma\eta}}(A_1^{\eta_1}A_2^{\eta_2}\cdots A_q^{\eta_q})^{1-\eta}.$$

PROOF. Changing indices if necessary, we may assume that

$$A_1 \geq A_2 \geq \cdots \geq A_q.$$

Then, the assumption implies that $M \geq A_2$. Set

$$\lambda_1 := \eta_1(1-\eta), \qquad \lambda_j := \frac{\eta_j}{\eta_2+\cdots+\eta_q}(1-\lambda_1) \quad (j = 2, \dots, q).$$

Since $\lambda_1 + \cdots + \lambda_q = 1$ and

$$\eta_j - \lambda_j \geq \eta_j(1-\eta) - \lambda_j = \frac{\eta_j(\gamma-1-\gamma\eta)}{\eta_2+\cdots+\eta_q} > 0$$

for every j, by (1.4.6) we have

$$\begin{aligned}
\sum_{j=1}^{q}\eta_jA_j &\geq \sum_{j=1}^{q}\lambda_jA_j \\
&\geq A_1^{\lambda_1}A_2^{\lambda_2}\cdots A_q^{\lambda_q} \\
&= (A_1^{\eta_1}A_2^{\eta_2}\cdots A_q^{\eta_q})^{1-\eta}\frac{A_2^{\lambda_2}\cdots A_q^{\lambda_q}}{(A_2^{\eta_2}\cdots A_q^{\eta_q})^{1-\eta}} \\
&= (A_1^{\eta_1}A_2^{\eta_2}\cdots A_q^{\eta_q})^{1-\eta}\frac{1}{A_2^{\eta_2(1-\eta)-\lambda_2}\cdots A_q^{\eta_q(1-\eta)-\lambda_q}} \\
&\geq (A_1^{\eta_1}A_2^{\eta_2}\cdots A_q^{\eta_q})^{1-\eta}\frac{1}{M^{\gamma-1-\gamma\eta}}.
\end{aligned}$$

This gives Lemma 1.4.7.

LEMMA 1.4.8. *For an arbitrarily given $\rho' > 0$ take a number a_0 ($> e$) satisfying the condition (1.4.4). Then, for each $\alpha \in \mathbf{C}$,*

$$\Delta \log \frac{1}{\log \dfrac{a_0}{|g,\alpha|^2}} \geq \frac{4|g'|^2}{(1+|g|^2)^2}\left(\frac{1}{|g,\alpha|^2 \log^2 \dfrac{a_0}{|g,\alpha|^2}} - \rho'\right). \tag{1.4.9}$$

PROOF. For brevity, we set

$$\varphi := |g,\alpha|^2 = \frac{|g-\alpha|^2}{(1+|g|^2)(1+|\alpha|^2)}.$$

Then, we have

$$\frac{\partial \varphi}{\partial z} = \frac{g'(\bar{g}-\bar{\alpha})(1+\alpha\bar{g})}{(1+|g|^2)^2(1+|\alpha|^2)}$$

and so

$$\left|\frac{\partial \varphi}{\partial z}\right|^2 = \frac{|g'|^2}{(1+|g|^2)^4}\frac{|g-\alpha|^2}{(1+|\alpha|^2)^2}|1+\bar{\alpha}g|^2.$$

Since

$$1-\varphi = \frac{|1+\bar{\alpha}g|^2}{(1+|g|^2)(1+|\alpha|^2)},$$

we obtain

$$\left|\frac{\partial \varphi}{\partial z}\right|^2 = (\varphi - \varphi^2)\frac{|g'|^2}{(1+|g|^2)^2}.$$

On the other hand, it holds that

$$\frac{\partial^2}{\partial z \partial \bar{z}} \log(1+|g|^2) = \frac{|g'|^2}{(1+|g|^2)^2}. \tag{1.4.10}$$

Therefore, we have

$$\begin{aligned}
\frac{1}{4}\Delta \log \frac{1}{\log(a/\varphi)} &= \frac{\partial}{\partial z}\left(\frac{1}{\log(a/\varphi)}\frac{\partial}{\partial \bar{z}}\log \varphi\right) \\
&= \frac{-1}{\log(a/\varphi)}\frac{\partial^2}{\partial z \partial \bar{z}}\log(1+|g|^2) + \frac{1}{\varphi^2 \log^2(a/\varphi)}\left|\frac{\partial \varphi}{\partial z}\right|^2 \\
&= \frac{|g'|^2}{(1+|g|^2)^2}\left(\frac{\varphi - \varphi^2}{\varphi^2 \log^2(a/\varphi)} - \frac{1}{\log(a/\varphi)}\right) \\
&= \frac{|g'|^2}{(1+|g|^2)^2}\left(\frac{1}{\varphi \log^2(a/\varphi)} - \left(\frac{1}{\log^2(a/\varphi)} + \frac{1}{\log(a/\varphi)}\right)\right).
\end{aligned}$$

If we take a constant $a := a_0(> 1)$ satisfying the condition (1.4.4), we have

$$\frac{1}{\log^2(a_0/\varphi)} + \frac{1}{\log(a_0/\varphi)} < \rho',$$

because $\varphi \leq 1$. Thus we get the desired inequality (1.4.9).

PROOF OF PROPOSITION 1.4.3. For brevity, we set

$$h_j := \frac{1}{|g, \alpha_j|} \qquad (1 \leq j \leq q).$$

Take $a_0(> e^2)$ satisfying the condition (1.4.4) for $\rho' := \rho/\gamma$. By Lemma 1.4.8 and (1.4.10) we see

$$\begin{aligned} &\Delta \log \frac{(1+|g|^2)^\rho}{\prod_{j=1}^q \log^{\eta_j}(a_0 h_j^2)} \\ (1.4.11) \qquad &\geq \frac{4|g'|^2}{(1+|g|^2)^2}\left(\rho + \sum_{j=1}^q \eta_j \left(\frac{h_j^2}{\log^2(a_0 h_j^2)} - \frac{\rho}{\gamma}\right)\right) \\ &= \frac{4|g'|^2}{(1+|g|^2)^2} \sum_{j=1}^q \frac{\eta_j h_j^2}{\log^2(a_0 h_j^2)}. \end{aligned}$$

On the other hand, for each $z \in \Delta_R$ it follows from (1.4.2) that $1 \geq |g(z), \alpha_j| \geq L/2$ for all except at most one α_j. Therefore, we have

$$\frac{h_j^2}{\log^2(a_0 h_j^2)} \leq \frac{4}{L^2 \log^2(4a_0/L^2)}$$

for such α_j's, because $x^2/\log^2(a_0 x^2)$ is monotone increasing for $x \geq 1$. Setting

$$M := \frac{4}{L^2 \log^2(4a_0/L^2)}, \quad A_j := \frac{h_j^2}{\log^2(a_0 h_j^2)} \qquad (1 \leq j \leq q),$$

we apply Lemma 1.4.7 to show that

$$\sum_{j=1}^q \frac{\eta_j h_j^2}{\log^2(a_0 h_j^2)} \geq \left(\frac{L}{2} \log \frac{4a_0}{L^2}\right)^{2(\gamma-1-\gamma\eta)} \prod_{j=1}^q \left(\frac{h_j^2}{\log^2(a_0 h_j^2)}\right)^{\eta_j(1-\eta)}.$$

In view of (1.4.11) this concludes Proposition 1.4.3.

We shall next prove the following:

PROPOSITION 1.4.12. *Let g be a nonconstant meromorphic function on Δ_R. Assume that, for some distinct points $\alpha_1, \alpha_2, \ldots, \alpha_q$ in $\bar{\mathbf{C}}$ and integers $m_1, m_2, \ldots, m_q$ not less than two, g does not take the values α_j with multiplicities less than m_j for each j and that*

$$\gamma := \sum_{j=1}^{q} \left(1 - \frac{1}{m_j}\right) > 2.$$

Then, for η_0 with $\gamma - 2 \geq \gamma\eta_0 > 0$ there is a constant $a_0 \geq e^2$ depending only on γ and η_0 such that, for an arbitrary nonnegative constant $\eta \leq \eta_0$, it holds that

$$\frac{|g'|}{1+|g|^2} \prod_{j=1}^{q} \left(\frac{1}{|g, \alpha_j| \log \dfrac{a_0}{|g, \alpha_j|^2}} \right)^{(1-1/m_j)(1-\eta)} \leq \frac{1}{C_1(1-\eta)^{1/2}} \frac{2R}{R^2 - |z|^2}$$

for the constant C_1 given by (1.4.5).

This will be proved by using the following Ahlfors-Schwarz lemma.

LEMMA 1.4.13 (cf., [2]). *If a continuous nonnegative function v on Δ_R is of class C^2 on the set $\{z \in \Delta_R; v(z) > 0\}$ and satisfies the condition*

$$\Delta \log v \geq v^2$$

there, then

$$v(z) \leq \frac{2R}{R^2 - |z|^2}$$

on Δ_R.

PROOF. Set

$$w_r(z) := \frac{2r}{r^2 - |z|^2}$$

for every $r > 0$. We can easily show

$$\Delta \log w_r = w_r^2. \tag{1.4.14}$$

For each $0 < r < R$ we consider the function

$$v_r(z) := \frac{v(z)}{w_r(z)} \equiv \frac{r^2 - |z|^2}{2r} v(z) \qquad (|z| \leq r).$$

Since $v_r(z) = 0$ on $\{z; |z| = r\}$, we can choose a point $z_0 \in \Delta_r$ such that

$$v_r(z_0) = \max\{v_r(z); |z| \leq r\}.$$

Assume that $v_r(z_0) > 1$. Then, the function $\log v_r$ takes the maximum at z_0, and so by the assumption and (1.4.14) we have

$$0 \geq \Delta \log v_r(z_0) = \Delta \log v(z_0) - \Delta \log w_r(z_0) \geq v(z_0)^2 - w_r(z_0)^2 > 0,$$

which is impossible. Consequently, we have $v_r(z) \leq 1$ for all $|z| < r$, namely,

$$v(z) \leq \frac{2r}{r^2 - |z|^2} \qquad (|z| < r).$$

By tending r to R, we obtain the desired inequality.

PROOF OF PROPOSITION 1.4.12. After a change of g by a suitable Möbius transformation, we may assume $\alpha_q = \infty$. Set

$$\eta_j := 1 - \frac{1}{m_j}, \quad h_j := \frac{1}{|g, \alpha_j|} \ (1 \leq j \leq q), \quad \rho := \frac{\gamma - 2 - \gamma\eta}{2(1 - \eta)}$$

for γ and η as in Proposition 1.4.12. Consider the function

$$v := C_1(1 - \eta)^{1/2} \frac{|g'|}{1 + |g|^2} \prod_{j=1}^{q} \left(\frac{h_j}{\log(a_0 h_j^2)} \right)^{\eta_j(1-\eta)}$$

where C_1 and a_0 are the constants as in Proposition 1.4.3 for the above γ, η $(\leq \eta_0)$ and

$$\rho' := \frac{\gamma - 2 - \gamma\eta_0}{2\gamma(1 - \eta_0)} \ (\leq \rho/\gamma).$$

Then, the exponent of the factor $(1 + |g|^2)^{1/2}$ appearing in v is equal to

$$(\eta_1 + \cdots + \eta_q)(1 - \eta) - 2 = \gamma(1 - \eta) - 2 = 2\rho(1 - \eta).$$

Setting

$$w := |g'| \prod_{j=1}^{q-1} \left(\frac{(1 + |\alpha_j|^2)^{1/2}}{|g - \alpha_j|} \right)^{\eta_j(1-\eta)},$$

we can rewrite v as

$$v = C_1(1-\eta)^{1/2} w \left(\frac{(1+|g|^2)^\rho}{\prod_{j=1}^q \log^{\eta_j}(a_0 h_j^2)} \right)^{1-\eta}$$

outside the set $E := \{z; g(z) \neq \alpha_j \text{for all j}\}$. Here, we claim that v is continuous on Δ_R and $\log w$ is harmonic on $\{z \in \Delta_R; w(z) > 0\}$ on $\Delta_R - E$. This is obviously true on $\Delta_R - E$. Consider first a point $z_0 \in \Delta_R$ with $g(z_0) = \alpha_j$ for some $j = 1, \ldots, q-1$, and let $m(\geq m_j)$ be the order of the zero of $g - \alpha_j$ at z_0. Write w as $w = |z - z_0|^a \tilde{w}$ with a positive function $\tilde{w}$ in some neighborhood of z_0. Then, since g' has a zero of order $m-1$ at z_0, we have

$$a = m - 1 - m\eta_j(1-\eta) = \frac{m}{m_j} - 1 + m\left(1 - \frac{1}{m_j}\right)\eta \geq 0.$$

Therefore, our claim is true around such a point z_0. Next, assume that g has a pole of order $m(\geq m_q)$ at a point z_0. Then, since g' has a pole of order $m+1$ at z_0, we can write w as $w = |z - z_0|^a \tilde{w}$ with a positive function $\tilde{w}$ around z_0 for

$$\begin{aligned} a &= -m - 1 + (\eta_1 + \cdots + \eta_{q-1})m(1-\eta) \\ &= m(\gamma - \gamma\eta - 2) + m - 1 - m\eta_q(1-\eta) \\ &\geq 2m\rho(1-\eta). \end{aligned}$$

On the other hand, the factor $(1+|g|^2)^{\rho(1-\eta)}$ in v is represented as a product of the function $|z-z_0|^{-2m\rho(1-\eta)}$ and a positive continuous function around z_0. Therefore, the function v is also continuous at z_0.

Now, we apply Proposition 1.4.3 to obtain

$$\begin{aligned} \Delta \log v &= (1-\eta)\Delta \log \left(\frac{(1+|g|^2)^\rho}{\prod_{j=1}^q \log^{\eta_j}(a_0 h_j^2)} \right) \\ &\geq (1-\eta) C_1^2 \frac{|g'|^2}{(1+|g|^2)^2} \prod_{j=1}^q \left(\frac{h_j^2}{\log^2(a_0 h_j^2)} \right)^{\eta_j(1-\eta)} \\ &= v^2. \end{aligned}$$

According to Lemma 1.4.13, we conclude Proposition 1.4.12.

COROLLARY 1.4.15. *Let g be a nonconstant meromorphic function on Δ_R satisfying the same assumptions as in Proposition 1.4.12. Then, for arbitrary constants $\eta \geq 0$ and $\delta > 0$ with $\gamma - 2 > \gamma\eta + \gamma\delta$, it holds that*

$$\frac{|g'|}{1+|g|^2} \frac{1}{\left(\prod_{j=1}^{q} |g, \alpha_j|^{1-1/m_j}\right)^{1-\eta-\delta}} \leq C_2 \frac{2R}{R^2 - |z|^2},$$

where C_2 is given by

$$C_2 := \frac{a_0^{\gamma\delta/2} C_3}{\delta^{\gamma(1-\eta)} \left(\dfrac{L}{2} \log \dfrac{4a_0}{L^2}\right)^{\gamma-1-\gamma\eta}}$$

for some constant C_3 depending only on γ.

PROOF. The function

$$\varphi(x) := \frac{\log^{1-\eta}(a_0 x^2)}{x^\delta} \qquad (1 \leq x < +\infty)$$

takes its maximum at a point $x_0 := \max\left\{\left(e^{2(1-\eta)/\delta}/a_0\right)^{1/2}, 1\right\}$. Therefore, we have

$$\begin{aligned}
&\frac{|g'|}{1+|g|^2} \frac{1}{\prod_{j=1}^{q} |g, \alpha_j|^{\eta_j(1-\eta-\delta)}} \\
&\quad = \frac{|g'|}{1+|g|^2} \prod_{j=1}^{q} \left(\frac{h_j}{\log(a_0 h_j^2)}\right)^{\eta_j(1-\eta)} \prod_{j=1}^{q} \left(\frac{\log(a_0 h_j^2)}{h_j^\delta}\right)^{\eta_j(1-\eta)} \\
&\quad \leq \frac{\varphi(x_0)^\gamma}{C_1(1-\eta)^{1/2}} \frac{2R}{R^2 - |z|^2}
\end{aligned}$$

by Proposition 1.4.12. Since $0 < \eta < (\gamma - 2)/\gamma$, we can find a positive constant C_3, depending only on γ, such that

$$\frac{\varphi(x_0)^\gamma}{C_1(1-\eta)^{1/2}} \leq \frac{2a_0^{\delta\gamma/2} C_3}{C_1 \delta^{\gamma(1-\eta)}}.$$

This concludes the proof of Corollary 1.4.15.

§1.5 The big Picard theorem

In this section, we give proofs of the big Picard theorem and related classical theorems for meromorphic functions as applications of the results in the previous section. The readers who are interested only in minimal surfaces may skip this section.

We give first the classical Landau's theorem.

THEOREM 1.5.1. *For arbitrarily given distinct numbers α and $\beta \neq 0$, we can find a positive constant $L(\alpha, \beta)$ such that, for any $R > L(\alpha, \beta)$, there is no nonconstant holomorphic function g on Δ_R satisfying the following condition:*
(1.5.2) $g(0) = \alpha,\ g'(0) = \beta$ *and* $g(z) \neq 0, 1$ *for each* $z \in \Delta_R$.

PROOF. Assume that there is a holomorphic function g on Δ_R for some positive R satisfying the condition (1.5.2). In this situation, setting $\alpha_1 := 0, \alpha_2 := 1$ and $\alpha_3 := \infty$ and considering $m_1 = m_2 = m_3 = \infty$, we can apply Corollary 1.4.15 to the function g. We can choose positive numbers η, δ such that, for some positive constant C,

$$\frac{|g'(z)|}{1 + |g(z)|^2} \frac{1}{\prod_{j=1}^{3} |g(z), \alpha_j|^{1-\eta-\delta}} \leq C \frac{2R}{R^2 - |z|^2}.$$

Substitute $z = 0$ into both sides of this inequality. We see easily

$$R \leq L(\alpha, \beta) := 2C \frac{1 + |\alpha|^2}{|\beta|} \prod_{j=1}^{3} |\alpha, \alpha_j|^{1-\eta-\delta}.$$

This proves Theorem 1.5.1.

We can easily conclude the following small Picard theorem from Theorem 1.5.1.

THEOREM 1.5.3. *Every nonconstant meromorphic function on $\mathbf{C}$ can omit at most two distinct values in $\bar{\mathbf{C}}$.*

PROOF. Suppose that a nonconstant meromorphic function g on $\mathbf{C}$ omits three distinct values. By using Möbius transformations if necessary, we may assume that $g'(0) \neq 0$ and the omitted three values are 0, 1 and ∞. Take some R with $R > L(g(0), g'(0))$ and consider the restriction of g to Δ_R. According to Theorem 1.5.1, there is no such a holomorphic function. This contradiction proves Theorem 1.5.3.

Next, we shall give a generalization of the classical Montel's theorem. We begin by giving the following definitions.

DEFINITION 1.5.4. Let $\{f_n; n = 1, 2, \ldots\}$ be a sequence of meromorphic functions on a domain D in $\mathbf{C}$. We shall say that $\{f_n\}$ is *compactly convergent in the wider sense* if every point in D has a neighborhood U satisfying the condition that $\{f_n\}$ converges uniformly on U or else there is some n_0 such that f_n has no zero for $n \geq n_0$ and $\{1/f_n; n = n_0, n_0+1, \dots\}$ converges uniformly on U.

DEFINITION 1.5.5. A family $\mathcal{F}$ of meromorphic functions on a domain D is called a *normal family in the wider sense* if every sequence in $\mathcal{F}$ has a subsequence which is compactly convergent in the wider sense.

THEOREM 1.5.6. *For a domain D in $\mathbf{C}$ and three distinct values α_1, α_2 and α_3 in $\bar{\mathbf{C}}$ let $\mathcal{F}(D; \alpha_1, \alpha_2, \alpha_3)$ be the family of all meromorphic functions f on D such that $f(z) \neq \alpha_j$ $(j = 1, 2, 3)$ for all $z \in D$. Then, $\mathcal{F}(D; \alpha_1, \alpha_2, \alpha_3)$ is a normal family in the wider sense.*

For the proof of Theorem 1.5.6, we shall first prove the following proposition under the same assumption as in Theorem 1.5.6:

PROPOSITION 1.5.7. *Let $\{g_n\}$ be a sequence in $\mathcal{F}(\Delta_R; \alpha_1, \alpha_2, \infty)$. If $\{g_n(0)\}$ has a limit in $\mathbf{C} - \{\alpha_1, \alpha_2\}$, $\{g_n\}$ has a subsequence which converges uniformly on every compact subset of Δ_R.*

PROOF(cf., [53, p. 94]). Consider a sequence $\{g_n\}$ as in Proposition 1.5.7 and take an arbitrary r_0 with $0 < r_0 < R$. For an arbitrary point z in Δ_{r_0} and arbitrary n, we set $g := g_n(z), g' := g'_n(z)$ and apply Proposition 1.4.12 to see

$$|g'| \leq \frac{1}{C_1} \frac{2R}{R^2 - r_0^2} (1 + |g|^2) \prod_{j=1}^{3} |g, \alpha_j| \log \frac{a_0}{|g, \alpha_j|^2}, \tag{1.5.8}$$

where we set $\alpha_3 := \infty$, $\eta := 0$ and $m_j := +\infty$ $(j = 1, 2, 3)$ in Proposition 1.4.12. On the other hand, by (1.4.2) we know $|g, \alpha_j| \geq L/2$ for all j except some j_0 with $|g, \alpha_{j_0}| = \min_{1 \leq j \leq 3} |g, \alpha_j|$. Therefore,

$$\prod_{j=1}^{3} \log \frac{a_0}{|g, \alpha_j|^2} \leq \log^2 \frac{4a_0}{L^2} \log \frac{a_0}{|g, \alpha_{j_0}|^2} \leq \log^2 \frac{4a_0}{L^2} \log \prod_{j=1}^{3} \frac{a_0}{|g, \alpha_j|^2}.$$

We consider the function

$$u(z) := \log \prod_{j=1}^{3} \frac{a_0}{|g, \alpha_j|^2},$$

which can be rewritten as

$$u(z) = \log \frac{a_0^3 (1 + |g|^2)^3 (1 + |\alpha_1|^2)(1 + |\alpha_2|^2)}{|g - \alpha_1|^2 |g - \alpha_2|^2}. \tag{1.5.9}$$

Using this function, we obtain from (1.5.8) as

$$|g'| \leq C_1'(1+|g|^2)u(z)\prod_{j=1}^{3}|g,\alpha_j| \qquad (z \in \Delta_{r_0}), \tag{1.5.10}$$

where C_1' is a constant depending only on R and r_0.

Now, for an arbitrarily fixed θ with $0 \leq \theta < 2\pi$, we define $v(r) := u(re^{i\theta})$. Since

$$\frac{dv}{dr} = \frac{\partial u}{\partial z}\frac{\partial z}{\partial r} + \frac{\partial u}{\partial \bar{z}}\frac{\partial \bar{z}}{\partial r} = \frac{\partial u}{\partial z}e^{i\theta} + \frac{\partial u}{\partial \bar{z}}e^{-i\theta},$$

we have

$$\left|\frac{dv}{dr}\right| \leq 2\left|\frac{\partial u}{\partial z}\right|.$$

On the other hand, by (1.5.9)

$$\frac{\partial u}{\partial z} = g'\left(\frac{3\bar{g}}{(1+|g|^2)} - \frac{1}{g-\alpha_1} - \frac{1}{g-\alpha_2}\right).$$

As is easily seen from this, there is a polynomial $P(X)$ of degree three with positive coefficients depending only on the α_j's such that

$$\left|\frac{\partial u}{\partial z}\right| \leq |g'|\frac{P(|g|)}{(1+|g|^2)|g-\alpha_1||g-\alpha_2|}.$$

We can find some $C_1'' > 0$ such that $P(|g|) \leq C_1''(1+|g|^2)^{3/2}$ and hence, with the help of (1.5.10), it follows that

$$\left|\frac{dv}{dr}\right| \leq |g'|\frac{C_1''}{(1+|g|^2)\prod_{j=1}^{3}|g,\alpha_j|} \leq C_1'C_1''v(r)$$

for some positive constant C_1''. In conclusion, there is a positive constant K_0 depending only on r_0 and α_j's such that

$$\left|\frac{d\log v(r)}{dr}\right| \leq K_0 \qquad (0 \leq r < r_0).$$

By integrating both sides, we see

$$\log v(r) \leq \log v(0) + K_0 r < \log v(0) + K_0 r_0$$

and so

$$v(r) \leq v(0)e^{K_0 r_0}. \tag{1.5.11}$$

Here, we recall the assumption that $\{g_n(0)\}$ has a limit in $\bar{\mathbf{C}} - \{\alpha_1, \alpha_2, \alpha_3\}$. The set of all values $v(0)$ of the functions v's corresponding to the functions g_n $(n = 1, 2, \ldots)$ is bounded. Hence, by (1.5.11) the set $\{v(r)\}$ is also bounded from above by a positive constant K_1 depending only on r_0 and α_j's. This implies that

$$(1+|g|^2)^{3/2} \leq K_1|g-\alpha_1||g-\alpha_2| \leq K_1((1+|\alpha_1|^2)(1+|\alpha_2|^2))^{1/2}(1+|g|^2).$$

Therefore, one concludes the uniform boundedness of $\{g_n(z); |z| < r_0, n = 1, 2, \ldots\}$. By the classical theorem on normal families, $\{g_n\}$ has a subsequence which converges on every compact subset of Δ_r. Since we can choose an arbitrary r_0 with $0 \leq r_0 < R$, $\{g_n\}$ has a subsequence converging uniformly on every compact subset of Δ_R. The proof of Proposition 1.5.7 is completed.

PROOF OF THEOREM 1.5.6. For our purpose, we may assume that $\alpha_3 = \infty$, and $D = \Delta_R$ $(R > 0)$ because we can use the usual diagonal method for general D. The proof is given by reduction to absurdity. Suppose that there is some sequence $\{f_n; n = 1, 2, \ldots\}$ in $\mathcal{F}(\Delta_R; \alpha_1, \alpha_2, \alpha_3)$ which has no subsequence converging compactly in the wider sense on Δ_R. Then, we may assume that there is some r with $0 < r < R$ such that any subsequence of $\{f_n\}$ is not compactly convergent on Δ_r. We set $U_j := \{w; |w, \alpha_j| < \delta\}$ $(j = 1, 2, 3)$ for a sufficiently small positive number δ so that $\bar{U}_j$'s are mutually disjoint.

Consider first the case where $f_n(\Delta_r) \subseteq \bigcup_{j=1}^q U_j$ for infinitely many n. For each of these n's the set $f_n(\Delta_r)$ is contained in some U_j because it is connected. Therefore, we can find a subsequence $\{f_{n_k}\}$ and some j_0 such that $f_{n_k}(\Delta_r) \subseteq U_{j_0}$ for all k. In this case, since the domain U_{j_0} is biholomorphic with a bounded domain in $\mathbf{C}$, $\{f_{n_k}\}$ has a subsequence which converge compactly in the wider sense on Δ_r by virtue of the classical theorem on normal families. This contradicts the assumption.

It remains to study the case where there is some n_0 such that $f_n(\Delta_r) \not\subseteq \bigcup_{j=1}^q U_j$ for all $n \geq n_0$. In this case, we can find a sequence $\{z_n; n = n_0, n_0+1, \ldots\}$ with $f_n(z_n) \notin \bigcup_j U_j$. Replacing it by a suitable subsequence, we may assume that $\{z_n\}$ and $\{f_n(z_n)\}$ have a limit z_0 in $\bar{\Delta}_r$ and a limit w_0 in $\bar{\mathbf{C}} - \bigcup_j U_j$ respectively. Consider the Möbius transformations

$$\psi_n(z) := \frac{R^2(z+z_n)}{R^2 + \bar{z}_n z}.$$

Now define

$$g_n(z) = f_n(\psi_n(z)).$$

Then, as is easily checked, $\{g_n(z)\}$ satisfies all assumptions in Proposition 1.5.7. Therefore, it has a subsequence which converges compactly in the wider sense. It follows that $\{f_n\}$ also has a subsequence which converges compactly in the wider sense, because $\{\psi_n^{-1}(z)\}$ converges compactly on Δ_R. This is a contradiction, which completes the proof of Theorem 1.5.6.

Now, we shall give a proof of the big Picard theorem.

THEOREM 1.5.12. *Let f be a meromorphic function on $\Delta^* := \{z; 0 < |z| < R\}$. If there are three distinct values $\alpha_1, \alpha_2, \alpha_3$ in $\bar{\mathbf{C}}$ such that each $f^{-1}(\alpha_j)$ $(j = 1, 2, 3)$ is finite, then f does not have an essential singularity at the origin.*

PROOF(cf., [53, p. 95]). Without loss of generality, we may assume that f omits the values $\alpha_j (j = 1, 2, 3)$ by shrinking R if necessary. The proof is given by reduction to absurdity. Suppose that f has an essential singularity at the origin. Then, by the classical Casorati-Weierstrass theorem on essential singularities there is a sequence $\{z_\nu\}$ in Δ^* such that $\lim_n z_n = 0$ and $\{f(z_n)\}$ converges also some value $\gamma \in \mathbf{C} - \{\alpha_1, \alpha_2\}$. By replacing $\{z_n\}$ by a suitable subsequence, we may assume that

$$Re^{-2\pi} > |z_1| > |z_2| > \cdots > |z_n| > \cdots \ .$$

Consider the functions

$$g_n(w) := f(z_n e^{2\pi i w}) \qquad (n = 1, 2, \dots)$$

of w on the open unit disc Δ_1. By the assumption, all g_n omit the values α_j's and

$$\lim_n g_n(0) = \lim_n f(z_n) = \gamma \in \mathbf{C} - \{\alpha_1, \alpha_2\}.$$

Proposition 1.5.7 yields that $\{g_n(z)\}$ has a subsequence which converges uniformly on every compact subset of Δ_1. There is no harm in assuming that $\{g_n\}$ itself converges compactly. Therefore, it converges uniformly on a particular set $E := \{w; w \in \mathbf{R}, -1/2 \leq w \leq 1/2\}$ and so there is a positive constant K such that $|g_n(w)| \leq K$ for all w in E. Since each point z with $|z| = |z_n|$ is represented as $z = |z_n| e^{2\pi i \theta}$ for some θ $(-1/2 \leq \theta \leq 1/2)$, we have $|f(z)| \leq K$ for all z with $|z| = |z_n|$. We can also conclude that $|f(z)| \leq K$ on the set $\{z; 0 < |z_{n+1}| \leq |z_n|\}$ for all n by the maximum principle. This shows that $|f|$ is uniformly bounded on the set $\{z; 0 < |z| \leq |z_1|\}$. By Riemann's theorem on removable singularities, f cannot have an essential singularity at the origin. This contradicts the assumption. We have Theorem 1.5.12.

§1.6 An estimate for the Gaussian curvature of minimal surfaces

Let $x : M \to \mathbf{R}^3$ be a minimal surface immersed in $\mathbf{R}^3$. A continuous curve $\gamma(t)$ $(0 \leq t \leq 1)$ in M is said to be *divergent* in M if, for each compact set K, there is some t_0 such that $\gamma(t) \notin K$ for any $t \geq t_0$. We define the distance $d(p)$ $(\leq +\infty)$ from a point $p \in M$ to the boundary of M as the greatest lower bound of the lengths of all continuous curves which are divergent in M.

The main purpose of this section is to prove the following:

THEOREM 1.6.1. *There exists a universal positive constant C with the following properties:*

Let $x : M \to \mathbf{R}^3$ be a minimal surface immersed in $\mathbf{R}^3$. Suppose that, for some fixed distinct values $\alpha_1, \ldots, \alpha_q$ and some integers $m_1, \ldots, m_q$ greater than one, the classical Gauss map g of M does not take the value α_j with multiplicity less than m_j for each j and that

$$\gamma := \sum_{j=1}^{q} \left(1 - \frac{1}{m_j}\right) > 4. \tag{1.6.2}$$

Set

$$L := \min\{|\alpha_i, \alpha_j|; 1 \leq i < j \leq q\}.$$

Then,

$$|K(p)|^{1/2} \leq \frac{C}{d(p)} \frac{\log^2 \frac{1}{L}}{L^3} \qquad (p \in M). \tag{1.6.3}$$

Before stating the proof of Theorem 1.6.1, we shall give some corollaries of it.

COROLLARY 1.6.4. *There exists a universal positive constant C with the following properties:*

Let $x = (x_1, x_2, x_3) : M \to \mathbf{R}^3$ be a nonflat minimal surface immersed in $\mathbf{R}^3$. Assume that the normals of M do not take five distinct directions $n_1, \ldots, n_5$, and set

$$L := \min\left\{\sin\left(\frac{\theta_{ij}}{2}\right); 1 \leq i < j \leq 5\right\}$$

for the angles θ_{ij} between n_i and n_j. Then,

$$|K(p)|^{1/2} \leq \frac{C}{d(p)} \frac{\log^2 \frac{1}{L}}{L^3} \qquad (p \in M). \tag{1.6.5}$$

PROOF. By the assumption, the classical Gauss map $g : M \to \bar{\mathbf{C}}$ omits the five distinct values $\alpha_j := \varpi(n_j)$ $(1 \leq j \leq 5)$, where ϖ denotes the stereographic projection of S^2 onto $\bar{\mathbf{C}}$. Setting $m_1 = \cdots = m_5 = \infty$, we apply Theorem 1.6.1 to the meromorphic function g. Corollary 1.6.4 is an immediate consequence of Theorem 1.6.1 because of (1.3.3).

By definition, a surface M is complete if and only if $d(p) = \infty$ for any point $p \in M$. We now give the following Picard type theorem for the classical Gauss map of minimal surfaces.

COROLLARY 1.6.6([36]). *The classical Gauss map of a complete non-flat minimal surface immersed in $\mathbf{R}^m$ can omit at most four distinct values.*

PROOF. If the classical Gauss map g omits five distinct values, the Gaussian curvature vanishes identically by virtue of Corollary 1.6.4 because $d(p) = \infty$ for all $p \in M$. This contradicts the assumption of non-flatness of M and so we conclude Corollary 1.6.6.

Now, we shall start to prove Theorem 1.6.1. We give first the following lemma.

LEMMA 1.6.7. *Let $d\sigma^2$ be a conformal flat metric on an open Riemann surface M. Then, for every point $p \in M$, there is a local diffeomorphism Φ of a disk $\Delta_{R_0} := \{w; |w| < R_0\}$ $(0 < R_0 \leq \infty)$ onto an open neighborhood of p with $\Phi(0) = p$ such that Φ is a local isometry, namely, the pull-back $\Phi^*(d\sigma^2)$ is equal to the standard metric on Δ_{R_0} and, for a point a_0 with $|a_0| = 1$, the Φ-image Γ_{a_0} of the curve*

$$L_{a_0} : w := a_0 s \qquad (0 \leq s < R)$$

is divergent in M.

PROOF. We may replace M and $d\sigma^2$ by the universal covering of the original M and the pull-back of the original $d\sigma^2$ respectively. Therefore, we may assume that M is simply connected. Then, by Koebe's theorem (e.g., [16, p. 210]), M is biholomorphic with the unit disc or the complex plane and so M has a global coordinate z. Represent $d\sigma^2$ as

$$d\sigma^2 = \lambda^2 |dz|^2,$$

where λ is a positive C^∞ function. By the assumption of flatness of $d\sigma^2$, $v := \log \lambda$ is a harmonic function. Since M is simply connected, we can take a harmonic function v^* on M such that $v + \sqrt{-1}v^*$ is holomorphic. Consider the holomorphic function $\psi = \exp(v + \sqrt{-1}v^*)$, which satisfies the condition $|\psi| = \lambda$. We define the holomorphic function

$$w = F(z) := \int_p^z \psi(\zeta)d\zeta, \tag{1.6.8}$$

where $\int_p^z$ means the integral along an arbitrarily chosen piecewise smooth curve joining a point p with z in M.

The function F satisfies the conditions $F(p) = 0$ and $dF(p) \neq 0$. Therefore, F maps an open neighborhood U of p biholomorphically onto an open disc $\Delta_R = \{w; |w| < R\}$ in $\mathbf{C}$, where $0 < R \leq +\infty$. Let R_0 be the least upper bound of $R > 0$ such that F biholomorphically maps some open neighborhood of p onto Δ_R. By definition, there is a sequence $\{R_n\}$ converging to R_0 such that $F|U_n : U_n \to \Delta_{R_n}$ is biholomorphic for open neighborhoods U_n of p. Then, F maps $U_0 := \cup_n U_n$ onto Δ_{R_0}. Consider the inverse map $\Phi := (F|U_0)^{-1}$ of $F|U_0$. We shall prove that $\Phi : \Delta_{R_0} \to U_0$ satisfies the desired condition.

For the case $R = \infty$, we have $M = U_0$, because an open Riemann surface which includes an open set biholomorphic with $\mathbf{C}$ is itself biholomorphic with $\mathbf{C}$. In this case, every curve Γ_{a_0} is divergent.

Consider the case $R < \infty$. Assume that the curve Γ_{a_0} is not divergent in M for some a_0 with $|a_0| = 1$. Then, there is some sequence $\{t_n\}$ with $0 \leq t_n < R_0$ such that $\lim_{n\to\infty} t_n = R_0$ and $\Phi(t_n a_0)$ converges to a point $q \in M$. Since $F(q) = \psi(q) \neq 0$, F biholomorphically maps some open neighborhood of q onto an open neighborhood of $w_0 := R_0 a_0$. Therefore, Φ is holomorphically extended to a neighborhood of w_0. If there exist no curve Γ_{a_0} which diverges in M, then Φ has a holomorphic extension to Δ_R for some $R > R_0$ and $F : \Phi(\Delta_R) \to \Delta_R$ is biholomorphic. This contradicts the definition of R_0. Therefore, some Γ_{a_0} diverges in M. Moreover, by (1.6.8) we have

$$\Phi^*(d\sigma^2) = (\lambda\circ\Phi)^2 \left|\frac{dz}{dw}\right|^2 |dw|^2 = e^{2v\circ\Phi}\frac{1}{|\psi|^2}|dw|^2 = |dw|^2.$$

This completes the proof of Lemma 1.6.7.

We note here that, for the proof of Theorem 1.6.1 there is no harm in assuming the following:

(1.6.9) (i) For any proper subset I of $\{1, 2, \dots, q\}$

$$\sum_{j \in I} \left(1 - \frac{1}{m_j}\right) \leq 4.$$

(ii) There is no set of positive integers $(m_1^*, \dots, m_q^*)$ distinct with $(m_1, \dots, m_q)$ satisfying the conditions

$$m_j^* \leq m_j \quad (1 \leq j \leq q), \quad \sum_{j=1}^{q} \left(1 - \frac{1}{m_j^*}\right) > 4. \tag{1.6.10}$$

For, if there is some proper subset I of $\{1, 2, \dots, q\}$ such that

$$\sum_{j \in I} \left(1 - \frac{1}{m_j}\right) > 4,$$

then we may use the values $\{\alpha_j; j \in I\}$ instead of the values $\{\alpha_1, \dots, \alpha_q\}$ given in Theorem 1.6.1. In fact, the assumptions of Theorem 1.6.1 hold for these values and, if (1.6.3) is valid for $\{\alpha_j; j \in I\}$, it is also valid for $\{\alpha_j; j = 1, \dots, q\}$. This shows that we may assume the condition (i) of (1.6.9) for the proof of Theorem 1.6.1. Moreover, if there is some $(m_1^*, \dots, m_q^*)$ satisfying the condition (1.6.10), we may prove Theorem 1.6.1 after replacing each integer m_j by m_j^*.

Next, we shall show the following:

LEMMA 1.6.11. *There are only finitely many sets of integers $m_1, \dots, m_q$ with $m_j \geq 2$ which satisfy the assumptions* (1.6.2) *and* (1.6.9).

PROOF. We take positive integers $m_1, \dots, m_q$ satisfying the conditions (1.6.2) and (1.6.9). Our task is to show that there are only finitely many choices of such integers. We may assume that

$$m_1 \leq \cdots \leq m_q.$$

Then, for the number

$$\gamma := \sum_{j=1}^{q} \left(1 - \frac{1}{m_j}\right)$$

we shall show

$$\gamma - 4 \le \frac{1}{m_q(m_q-1)}. \tag{1.6.12}$$

To this end, suppose that $\gamma - 4 > 1/m_q(m_q-1)$. If $m_q = 2$, then $\gamma > 4+1/2$ and so $\sum_{j=1}^{q-1}(1-1/m_j) > 4$, which contradicts the assumption (i) of (1.6.9). Therefore, $m_q \ge 3$. Here, if we set $m_j^* := m_j$ $(1 \le j \le q-1)$ and $m_q^* := m_q - 1$, then

$$\sum_{j=1}^{q}\left(1-\frac{1}{m_j^*}\right) = \sum_{j=1}^{q}\left(1-\frac{1}{m_j}\right) - \frac{1}{m_q(m_q-1)} > 4.$$

This contradicts the assumption (ii) of (1.6.9). Therefore, we conclude (1.6.12).

By virtue of (1.6.12), we get

$$\gamma \le 4 + \frac{1}{m_q(m_q-1)} < 4 + \frac{1}{2} = \frac{9}{2}.$$

On the other hand, since $m_j \ge 2$ for all j, we have

$$\gamma = \sum_j \left(1-\frac{1}{m_j}\right) \ge q\left(1-\frac{1}{m_1}\right) \ge \frac{q}{2},$$

where $q \ge 5$. Therefore, we conclude $m_1 < 2q/(2q-9)$ and $q < 9$.

Now, by induction on k $(= 1, \dots, q)$ we shall show that, for preassigned $m_1, \dots, m_k$, the number m_{k+1} among $m_1, \dots, m_k, m_{k+1}, \dots, m_q$ satisfying the desired conditions is bounded from above by a constant depending only on q and $m_1, \dots, m_k$, from which we can easily conclude Lemma 1.6.11. The boundedness of m_1 has been already shown. Consider the numbers $m_1, \dots, m_q$ satisfying the desired conditions. Set

$$\delta_0 := \sum_{j=1}^{k}\left(1-\frac{1}{m_j}\right).$$

Then, since

$$4 < \gamma = \delta_0 + \sum_{j=k+1}^{q}\left(1-\frac{1}{m_j}\right),$$

we have

$$\delta_0 + q - k - 4 = \gamma - 4 + \sum_{j=k+1}^{q} \frac{1}{m_j} > 0.$$

Take a number A with $\delta_0 + q - k - 4 > \eta_0 := 1/A(A-1)$. If $m_q \leq A$, then we have $m_{k+1} \leq m^*$. Otherwise, by the use of (1.6.12) and the inequalities $m_k \leq m_j$ for $j = k+1, \dots, q$, we have

$$\begin{aligned} 0 < \delta_0 + q - k - \eta_0 - 4 &\leq \delta_0 + q - k - 4 - \frac{1}{m_q(m_q - 1)} \\ &\leq \sum_{j=k+1}^{q} \frac{1}{m_j} \leq \frac{q-k}{m_{k+1}}. \end{aligned}$$

This gives

$$m_{k+1} \leq \frac{q-k}{\delta_0 - 4 + q - k - \eta_0}.$$

In conclusion,

$$m_{k+1} \leq \max\left\{A, \frac{q-k}{\delta_0 + q - k - \eta_0}\right\}.$$

This completes the proof of Lemma 1.6.11.

With the help of Lemma 1.6.11, if we take the maximum C_0 of constants C which are chosen for the finitely many possible cases of m_j's satisfying the condition (1.6.2) and (1.6.9), then C_0 satisfies the desired inequality (1.6.3). For the proof of Theorem 1.6.1 it suffices to show the existence of a constant satisfying (1.6.3) which may depend on the given data $m_1, \dots, m_q$.

We consider a minimal surface $x := (x_1, x_2, x_3) : M \to \mathbf{R}^3$ immersed in $\mathbf{R}^3$ whose classical Gauss map $g : M \to \bar{\mathbf{C}}$ satisfies the assumption of Theorem 1.6.1 for $\alpha_1, \dots, \alpha_q$ and integers $m_1, \dots, m_q$ with $m_j \geq 2$. We may assume that M is non-flat, or equivalently that g is not a constant, because otherwise Theorem 1.6.1 is trivial. Moreover, we may assume $\alpha_q = \infty$ after a suitable Möbius transformation of $\mathbf{C}$.

Taking a holomorphic local coordinate z, set $f_i := \partial x_i / \partial z$ $(i = 1, 2, 3)$. Then, by (1.3.6) we have $g = f_3/(f_1 - \sqrt{-1} f_2)$ and the induced metric on M is given by $ds^2 = |h_z|^2 (1 + |g|^2)^2 |dz|^2$ for the holomorphic function $h_z := f_1 - \sqrt{-1} f_2$, where h_z has a zero of order $2k$ at each point where g has a pole of order k.

Now, we choose some δ such that

$$\gamma - 4 > 2\gamma\delta > 0 \tag{1.6.13}$$

and set

$$\eta := \frac{\gamma - 4 - 2\gamma\delta}{\gamma}, \qquad \tau := \frac{2}{2 + \gamma\delta}. \tag{1.6.14}$$

Then, if we choose a sufficiently small positive δ depending only on γ, for the constant $\varepsilon_0 := (\gamma - 4)/2\gamma$ we have

$$0 < \tau < 1, \quad \frac{\varepsilon_0 \tau}{1 - \tau} \left(= \frac{\gamma - 4}{\delta\gamma^2} \right) > 1. \tag{1.6.15}$$

We consider the function

$$\xi_z := |h_z|^{1/(1-\tau)} \left(\frac{1}{|g'_z|} \prod_{j=1}^{q-1} \left(\frac{|g - \alpha_j|}{(1 + |\alpha_j|^2)^{1/2}} \right)^{\eta_j(1-\eta-\delta)} \right)^{\tau/(1-\tau)} \tag{1.6.16}$$

defined on the set

$$M' := \{p \in M; g'_z(p) \neq 0 \text{ and } g(p) \neq \alpha_j \text{ for all } j\},$$

where $\eta_j := 1 - 1/m_j$ and g'_z denotes the derivative of g with respect to the holomorphic local coordinate z. If we choose another holomorphic local coordinate ζ, then ξ_z is multiplied by $|d\zeta/dz|$. This shows that

$$d\sigma^2 := \xi_z^2 |dz|^2$$

is a well-defined metric on M'.

Our purpose is to show the inequality (1.6.3) for each point $p \in M$. We may assume that $p \in M'$, because $K(p)$ is continuous and $d(p)$ is lower semi-continuous. As is easily seen, the metric $d\sigma^2$ is flat on M'. According to Lemma 1.6.7, there is a local isometry Φ of a disk $\Delta_R := \{w; |w| < R\}$ $(0 < R \leq \infty)$ onto an open set in M, with respect to the standard metric and the metric $d\sigma^2$ respectively, such that, for a point a_0 with $|a_0| = 1$, the Φ-image Γ of the curve $L_{a_0} : w := a_0 s$ $(0 \leq s < R)$ is divergent in M'. For brevity, we denote the function $g \circ \Phi$ on Δ_R by g in the following. According to Corollary 1.4.15, we have

$$R \leq 2C_2 \frac{1 + |g(0)|^2}{|g'_z(0)|} \prod_{j=1}^{q} |g(0), \alpha_j|^{\eta_j(1-\eta-\delta)} < +\infty \tag{1.6.17}$$

for the constant C_2 given in Corollary 1.4.15. Therefore,

$$L_{d\sigma}(\Gamma) = \int_\Gamma d\sigma = \int_{L_{a_0}} = R < \infty,$$

where $L_{d\sigma}(\Gamma)$ denotes the length of Γ with respect to the metric $d\sigma^2$.

Now, suppose that the image $\gamma := \Phi(\Gamma)$ of Γ is not divergent in M. Then, since γ is divergent in M' and $L_{d\sigma}(\Gamma) < +\infty$, it must tend to a point p_0 where $g'(p_0) = 0$ or $g(p_0) = \alpha_j$ for some j. Taking a holomorphic local coordinate ζ in a neighborhood of p_0 with $\zeta(p_0) = 0$, we write the metric $d\sigma^2$ as $d\sigma^2 = |\zeta|^{2a\tau/(1-\tau)}w|d\zeta|^2$ with some positive C^∞ function w and some real number a. If $g - \alpha_j$ has a zero of order $m(\geq m_j \geq 2)$ at p_0 for some $j \leq q-1$, then g'_z has a zero of order $m-1$ at p_0 and $h_z(p_0) \neq 0$. Then,

$$\begin{aligned} a &= m\left(1 - \frac{1}{m_j}\right)(1-\eta-\delta) - (m-1) \\ &= 1 - \frac{m}{m_j} - \frac{m}{m_j}(m_j - 1)(\eta+\delta) \\ &\leq -(\eta+\delta) \leq -\varepsilon_0. \end{aligned}$$

For the case where g has a pole of order $m(\geq m_q)$ at p_0, g'_z has a pole of order $m+1$, h_z has a zero of order $2m$ at p_0 and each factor $g - \alpha_j$ in the right hand side of (1.6.16) has a pole of order m. Using the identity $\eta_1 + \cdots + \eta_{q-1} = \gamma - \eta_q$ and (1.6.14) we have

$$\begin{aligned} a &= \frac{2m}{\tau} + m + 1 - m(\gamma - \eta_q)(1-\eta-\delta) \\ &= m(2+\gamma\delta) + m + 1 - m\gamma + m(\gamma - 4 - \gamma\delta) + m\eta_q(1-\eta-\delta) \\ &= m\eta_q(1-\eta-\delta) - (m-1) \\ &\leq -\varepsilon_0. \end{aligned}$$

Moreover, for the case where $g'_z(p_0) = 0$ and $g(p_0) \neq \alpha_j$ for all j, then we see $a \leq -1$. In any case, $a\tau/(1-\tau) \leq -1$ by the use of (1.6.15), and there is a positive constant C_4 such that

$$d\sigma \geq C_4 \frac{|d\zeta|}{|\zeta|}$$

in a neighborhood of p_0.

$$R = \int_\Gamma d\sigma \geq C_4 \int_\Gamma \frac{1}{|\zeta|}|d\zeta| = +\infty,$$

which contradicts (1.6.17). Therefore, γ diverges in M as t tends to 1.

To estimate the length of γ, we shall study the metric $\Phi^* ds^2$ on Δ_R. For local considerations, the coordinate z on Δ_R may be considered as a holomorphic local coordinate on M' and so we may write $d\sigma^2 = |dz|^2$. By (1.6.16) we obtain

$$1 = |h_z|^{2/(1-\tau)} \left(\frac{1}{|g'_z|} \prod_{j=1}^{q-1} \left(\frac{|g-\alpha_j|}{(1+|\alpha_j|^2)^{1/2}} \right)^{\eta_j(1-\eta-\delta)} \right)^{2\tau/(1-\tau)}$$

and hence

$$|h_z| = \left(|g'_z| \prod_{j=1}^{q-1} \left(\frac{(1+|\alpha_j|^2)^{1/2}}{|g-\alpha_j|} \right)^{\eta_j(1-\eta-\delta)} \right)^{\tau}. \tag{1.6.18}$$

By the use of Corollary 1.4.15 we have

$$\begin{aligned} \Phi^* ds &= |h_z|(1+|g|^2)|dz| \\ &= \left(|g'_z|(1+|g|^2)^{1/\tau} \prod_{j=1}^{q-1} \left(\frac{(1+|\alpha_j|^2)^{1/2}}{|g-\alpha_j|} \right)^{\eta_j(1-\eta-\delta)} \right)^{\tau} |dz| \\ &= \left(\frac{|g'_z|}{1+|g|^2} \frac{1}{\prod_{j=1}^{q} |g, \alpha_j|^{\eta_j(1-\eta-\delta)}} \right)^{\tau} |dz| \\ &\leq C_2^{\tau} \left(\frac{2R}{R^2-|z|^2} \right)^{\tau} |dz|. \end{aligned}$$

This yields that

$$\begin{aligned} d(p) \leq \int_{\gamma} ds = \int_{\Gamma} \Phi^* ds \leq C_2^{\tau} \int_{\Gamma} \left(\frac{2R}{R^2-|z|^2} \right)^{\tau} |dz| \\ = C_2^{\tau} \int_0^R \left(\frac{2R}{R^2-x^2} \right)^{\tau} dx \leq \frac{(2C_2)^{\tau} R^{1-\tau}}{1-\tau}. \end{aligned}$$

By (1.6.17) we obtain

$$d(p) \leq \frac{2C_2}{1-\tau} \left(\frac{(1+|g(0)|^2) \prod_{j=1}^{q} |g(0), \alpha_j|^{\eta_j(1-\eta-\delta)}}{|g'_z(0)|} \right)^{1-\tau}.$$

On the other hand, in view of (1.6.18) the curvature at p is given by

$$\begin{aligned} |K|^{1/2} &= \frac{|g'_z|}{|h_z|(1+|g|^2)^2} \\ &= \frac{|g'_z|}{(1+|g|^2)^2}\left(\frac{(1+|g|^2)^{\gamma(1-\eta-\delta)/2}\prod_{j=1}^q |g,\alpha_j|^{\eta_j(1-\eta-\delta)}}{|g'_z|}\right)^{\tau}. \end{aligned}$$

Since $|g,\alpha_j| \leq 1$, we can easily conclude that

$$|K(p)|^{1/2} d(p) \leq C_5 := \frac{2C_2}{1-\tau}.$$

By the definition of C_2 and τ, we see

$$C_5 = \frac{2a_0^{\gamma\delta/2} C_3(2+\gamma\delta)}{\delta^{\gamma(1-\eta)}\gamma\delta\left(\dfrac{L}{2}\log\dfrac{4a_0}{L^2}\right)^{\gamma-1-\gamma\eta}}.$$

Now, take a sufficiently small L_0 satisfying the condition

$$\frac{1}{\log\frac{4a_0}{L_0^2}} < \frac{\gamma-4}{2\gamma}.$$

For each positive $L(\leq 1)$ we set $\delta := 1/\log(4a_0/L^2)$ if $L \leq L_0$ and $\delta := \delta_0$ for some δ_0 with $0 < \delta_0 < (\gamma-4)/2\gamma$ if $L_0 < L \leq 1$. Since a_0 can be chosen so as to be between two positive constants depending only on γ, we can conclude

$$C_5 \leq \frac{C_6 \log^2 \dfrac{4a_0}{L^2}}{L^{\gamma-1-\gamma\eta}} \leq C_7 \frac{\log^2 \dfrac{1}{L}}{L^3 L^{2\gamma\delta}},$$

for positive constants C_6 and C_7 depending only on γ. On the other hand, the factor

$$L^{2\gamma\delta} = \exp\left(2\gamma \log L/\log(4a_0/L^2)\right)$$

is bounded from below by a positive constant not depending on each L. This shows that C_7 can be replaced by a positive constant depending only on the m_j. The proof of Theorem 1.6.1 is complete.

It is an interesting open problem to determine whether the factor

$$\varphi(L) := \frac{1}{L^3}\log^2\left(\frac{1}{L}\right)$$

in the inequality of Corollary 1.6.4 can be replaced by $1/L^3$ or not. Related to this, we note that $\varphi(L)$ cannot be replaced by a number smaller than $1/L^{3-\varepsilon}$ for any $\varepsilon > 0$. In fact, for an arbitrarily given $\varepsilon > 0$ we can give an example of a family of minimal surfaces $\{M_\iota\}$ such that there is no positive constant C, not depending on M_ι, satisfying the condition

$$|K(p)|^{1/2} \leq \frac{C}{d(p)} \frac{1}{L^{3-\varepsilon}} \tag{1.6.19}$$

for every M_ι. To see this, for each positive number $R(\geq 1)$ we take five points

$$\alpha_1 := R,\ \alpha_2 := \sqrt{-1}R,\ \alpha_3 := -R,\ \alpha_4 := -\sqrt{-1}R,\ \alpha_5 := \infty$$

in $\bar{\mathbf{C}}$. Consider the form $\omega = dz$ and $g(z) = z$ on the disc $\Delta_R := \{z; |z| < R\}$. These functions satisfies the assumption of Theorem 1.3.12 and the functions

$$\omega_1 = \frac{1}{2}(1 - z^2)dz,\ \omega_2 = \frac{\sqrt{-1}}{2}(1 + z^2)dz,\ \omega_3 = zdz$$

have obviously no real period. Setting

$$x_i := 2 \operatorname{Re} \int_0^z \omega_i \qquad (1 \leq i \leq 3),$$

we can define a minimal surface $x = (x_1, x_2, x_3) : M_R := \Delta_R \to \mathbf{R}^3$ immersed in $\mathbf{R}^3$ whose classical Gauss map is the function g and whose induced metric is given by $ds^2 = (1 + |z|^2)^2 |dz|^2$. This is a portion of the minimal surface called Enneper surface. Consider the quantities $K(p)$ and $d(p)$ as in Corollary 1.6.4 at the point $p = 0$. We have

$$d(0) = \int_0^R (1 + x^2)dx = R + \frac{1}{3}R^3$$

and

$$|K(0)|^{1/2} = \frac{2|g'(0)|}{(1 + |g(0)|^2)^2} = 2.$$

On the other hand, the quantity L for the points in S^2 corresponding to the α_j's is given by $L = 1/\sqrt{1 + R^2}$ and so

$$|K(0)|^{1/2} d(0) L^{3-\varepsilon} = \frac{2(R + \frac{1}{3}R^3)}{(1 + R^2)^{(3-\varepsilon)/2}},$$

which converges to $+\infty$ as R tends to $+\infty$. Therefore, there is no positive constant satisfying the condition (1.6.19) and not depending on each minimal surface M_R.

Chapter 2

The derived curves of a holomorphic curve

§2.1 Holomorphic curves and their derived curves

Let M be an open Riemann surface. By a divisor on M we mean a map ν of M into $\mathbf{R}$ whose support $\mathrm{Supp}(\nu) := \overline{\{p \in M; \nu(p) \neq 0\}}$ has no accumulation points in M. We begin by explaining the divisor associated with a function on M possibly with singularities in a discrete subset of M.

We shall call u a function with *mild singularities* on a domain D if u is a complex-valued function of class C^∞ on D outside a discrete set E and every point $a \in D$ has a neighborhood U such that for a holomorphic local coordinate z with $z(a) = 0$ on U we can write

$$|u(z)| = |z|^\sigma u^*(z) \prod_{j=1}^{q} \left| \log \frac{1}{|g_j(z)| v_j(z)} \right|^{\tau_j} \tag{2.1.1}$$

on $U - E$ with some real number σ, nonpositive real numbers τ_j, nonzero holomorphic functions g_j with $g_j(0) = 0$ and some positive C^∞ functions u^* and v_j on U, where $0 \leq q < \infty$.

For a function u with mild singularities on a domain D, we define the map $\nu_u : D \to \mathbf{R}$ of u by

$$\nu_u(a) := \text{the number } \sigma \text{ appearing in a representation (2.1.1) of } u$$

for each $a \in D$, which is well defined on D because it does not depend on a choice of a holomorphic local coordinate z. We call ν_u the divisor of u.

The product of two functions u_1 and u_2 with mild singularities is also a function with mild singularities and we have

$$\nu_{u_1 u_2} = \nu_{u_1} + \nu_{u_2}.$$

For a nonzero meromorphic function f, the divisor ν_f of f is nothing but the function whose value of each point a is the order of f at a, namely,

$$\nu_f(a) = \begin{cases} m & \text{if } f \text{ has a zero of order } m \text{ at } a, \\ -p & \text{if } f \text{ has a pole of order } p \text{ at } a, \\ 0 & \text{otherwise.} \end{cases}$$

As is well-known(e.g., [16, Theorem 26.5]), it holds that

(2.1.2) *For every integer-valued divisor ν on M, there exists a nonzero meromorphic function f on M such that $\nu = \nu_f$.*

Let f be a holomorphic curve in $P^n(\mathbf{C})$ defined on an open Riemann surface M, which means a nonconstant holomorphic map of M into $P^n(\mathbf{C})$. For a fixed system of homogeneous coordinates $(w_0 : \cdots : w_n)$ we set

$$V_i = \{(w_0 : \cdots : w_n); w_i \neq 0\} \qquad (1 \leq i \leq q).$$

Then, every $a \in M$ has a neighborhood U of a such that $f(U) \subset V_i$ for some i and f has a representation

$$f = (f_0 : \cdots : f_{i-1} : 1 : f_{i+1} : \cdots : f_n)$$

on U with holomorphic functions $f_0, \ldots, f_{i-1}, f_{i+1}, \ldots, f_n$.

DEFINITION 2.1.3. For an open subset U of M we call a representation $f = (f_0 : \cdots : f_n)$ to be a *reduced representation* of f on U if $f_0, \ldots, f_n$ are holomorphic functions on U and have no common zero.

As is stated above, every holomorphic map of M into $P^n(\mathbf{C})$ has a reduced representation on some neighborhood of each point in M. Moreover, we see easily

(2.1.4) *Let $f = (f_0 : \cdots : f_n)$ be a reduced representation of f. Then, for an arbitrary nowhere zero holomorphic function h, $f = (f_0 h : \cdots : f_n h)$ is also a reduced representation of f. Conversely, for every reduced representation $f = (g_0 : \cdots : g_n)$ of f, each g_i can be written as $g_i = hf_i$ with a nowhere zero holomorphic function h.*

We now take $n+1$ holomorphic functions $f_0, \ldots, f_n$ on M at least one of which does not vanish identically. Let

$$\nu_0(a) := \min\{\nu_{f_i}(a); f_i \not\equiv 0, 0 \leq i \leq n\} \qquad (a \in M),$$

and take a nonzero holomorphic function g with $\nu_g = \nu_0$. Then, $f_i/g\,(0 \leq i \leq n)$ are holomorphic functions without common zeros. We can define a holomorphic map f with a reduced representation $f = (f_0/g : \cdots : f_n/g)$, which we call the holomorphic curve defined by $f_0, \ldots, f_n$.

Take a holomorphic map f of M into $P^n(\mathbf{C})$ and a hyperplane H in $P^n(\mathbf{C})$ not including the image $f(M)$ of f. The hyperplane H can be written as

$$H:\ (W, A) \equiv a_0 w_0 + \cdots + a_n w_n = 0$$

with a nonzero vector $A = (a_0, \dots, a_n) \in \mathbf{C}^{n+1}$. For each point $a \in M$ choosing a reduced representation $f = (f_0 : \cdots : f_n)$ on a neighborhood U of a, we consider the holomorphic function $F(H) := a_0 f_0 + \cdots + a_n f_n$ on U and set $\nu(f, H) := \nu_{F(H)}$. Since $\nu(f, H)$ depends only on f and H, we can define the divisor $\nu(f, H)$ on the totality of M, which we call the *pull-back* of H considered as a divisor.

PROPOSITION 2.1.5. *Every holomorphic map f of an open Riemann surface M into $P^n(\mathbf{C})$ has a reduced representation on the totality of M.*

PROOF. Set $H_i := \{w_i = 0\}$ $(0 \leq i \leq n)$ for a fixed system of homogeneous coordinates $(w_0 : \cdots : w_n)$. Changing indices if necessary, we may assume that $f(M) \not\subset H_0$ and so $\nu(f, H_0)$ is well-defined. By (2.1.2) there is a nonzero holomorphic function g such that $\nu_g = \nu(f, H_0)$. On the other hand, we can take an open covering $\{U_\kappa; \kappa \in I\}$ of M such that f has a reduced representation $f = (f_{\kappa 0} : \cdots : f_{\kappa n})$ on each U_κ. Then, $\nu_g = \nu_{f_{\kappa 0}}$. Set $g_{\kappa i} := (g/f_{\kappa 0}) f_{\kappa i}$ $(0 \leq \kappa \leq n)$, which is holomorphic on U_κ. If $U_{\kappa\lambda} := U_\kappa \cap U_\lambda \neq \phi$, then there is a nowhere zero holomorphic function h with $f_{\lambda i} = h f_{\kappa i}$ on $U_{\kappa\lambda} (0 \leq i \leq n)$ by (2.1.4). We have

$$g_{\kappa i} = \frac{h f_{\kappa i} g}{h f_{\kappa 0}} = \frac{f_{\lambda i} g}{f_{\lambda 0}} = g_{\lambda i}$$

on $U_{\kappa\lambda}$ for each i. Therefore, we can define the function g_i on the totality of M which equals $g_{\kappa i}$ on each U_κ. For these functions, $f = (g_0 : \cdots : g_n)$ is a reduced representation of M.

Now, we consider k arbitrarily given holomorphic functions $f_0, \dots, f_k$ on M. For a holomorphic local coordinate z on an open subset U of M, we denote by $(f_i^{(\ell)})_z$, or simply by $f_i^{(\ell)}$, the ℓ-th derivative of f_i with respect to z, where we set $(f_i^{(0)})_z := f_i$. By definition, the Wronskian of $f_0, \dots, f_k$ is given by

$$W(f_0, \dots, f_k) \equiv W_z(f_0, \dots, f_k) := \det\left((f_i^{(\ell)})_z; 0 \leq i, \ell \leq k\right).$$

PROPOSITION 2.1.6. *For two holomorphic local coordinates z and ζ, it holds that*

$$W_\zeta(f_0, \dots, f_k) = W_z(f_0, \dots, f_k) \left(\frac{dz}{d\zeta}\right)^{k(k+1)/2}.$$

PROOF. Set $F = (f_0, \dots, f_k)$ and $(F^{(\ell)})_z = ((f_0^{(\ell)})_z, \dots, (f_k^{(\ell)})_z)$. By induction on ℓ it is easily seen that there are suitable polynomials $g_{\ell m}$ in $\dfrac{dz}{d\zeta}, \dots, \dfrac{d^\ell z}{d\zeta^\ell}$ such that

$$(F^{(\ell)})_\zeta = (F^{(\ell)})_z \left(\frac{dz}{d\zeta}\right)^\ell + \sum_{m=0}^{\ell-1} g_{\ell m}(F^{(m)})_z.$$

Substitute these identities into the formula

$$W_\zeta(f_0, \dots, f_k) = \det({}^t(F^{(0)})_\zeta, \dots, {}^t(F^{(k)})_\zeta)$$

and subtract the first column multiplied by g_{01} from the second column to see

$$W_\zeta(f_0, \dots, f_k) = \det({}^t(F^{(0)})_z, {}^t(F^{(1)})_z \frac{dz}{d\zeta}, {}^t(F^{(2)})_\zeta, \dots, {}^t(F^{(k)})_\zeta),$$

where ${}^tF^{(\ell)}$ denotes the transpose of the vector $F^{(\ell)}$. One may then subtract the sum of the first column multiplied by g_{02} and the second column multiplied by g_{12} from the third column. We repeat these processes and obtain

$$\begin{aligned} W_\zeta(f_0, \dots, f_k) &= \det \left({}^t(F^{(0)})_z, {}^t(F^{(1)})_z \frac{dz}{d\zeta}, \dots, {}^t(F^{(k)})_z \left(\frac{dz}{d\zeta}\right)^k\right) \\ &= \det({}^t(F^{(0)})_z, {}^t(F^{(1)})_z, \dots, {}^t(F^{(k)})_z) \left(\frac{dz}{d\zeta}\right)^{k(k+1)/2}. \end{aligned}$$

This concludes Proposition 2.1.6.

PROPOSITION 2.1.7. *For holomorphic functions $f_0, \dots, f_k$ on M the following conditions are equivalent:*

(i) *$f_0, \dots, f_k$ are linearly dependent over $\mathbf{C}$.*

(ii) *$W_z(f_0, \dots, f_k) \equiv 0$ for some (or all) holomorphic local coordinate z.*

PROOF. We first assume that $f_0, \dots, f_k$ are linearly dependent. Then the vector-valued function $F = (f_0, \dots, f_k)$ satisfies the identity $(A, F) = 0$ for a non-zero vector A and we have $(A, F^{(\ell)}) = 0$ for $\ell = 0, \dots, k$, whence the condition (ii) is satisfied for any holomorphic local coordinate

z. Here, by the theorem of identity, the condition (ii) is satisfied for every holomorphic local coordinate z if it is satisfied for some holomorphic local coordinate.

We next assume that the condition (ii) is satisfied. By $\mathcal{M}$ we denote the field of all meromorphic functions on an open set U on which a holomorphic local coordinate z is defined, and consider the $(k+1)$-dimensional vector space $\mathcal{F} := \mathcal{M}^{k+1}$ over $\mathcal{M}$. As in the case of a vector space over $\mathbf{C}$, vectors $G_0, G_1, \dots, G_r$ in $\mathcal{F}$ are linearly independent if and only if $\operatorname{rank}({}^tG_0, \dots, {}^tG_r) = r+1$, namely, some minor of the matrix $({}^tG_0, \dots, {}^tG_r)$ of degree $r+1$ does not vanish identically. The condition (ii) yields that there is some m with $0 \leq m \leq k$ such that $F^{(m)}$ can be written as a linear combination of $F, F^{(1)}, \dots, F^{(m-1)}$ over $\mathcal{M}$. Then, by induction on ℓ we can show that all $F^{(\ell)}$ are written as linear combinations of $F, F^{(1)}, \dots, F^{(m)}$. In fact, if $F^{(\ell)}$ is written as

$$F^{(\ell)} = a_{\ell 0}F^{(0)} + a_{\ell 1}F^{(1)} + \cdots + a_{\ell m-1}F^{(m-1)}$$

for some $a_{\ell j} \in \mathcal{M}$, then

$$\begin{aligned} F^{(\ell+1)} = {} & a'_{\ell 0}F^{(0)} + a'_{\ell 1}F^{(1)} + \cdots + a'_{\ell m-1}F^{(m-1)} \\ & + a_{\ell 0}F^{(1)} + \cdots + a_{\ell m-2}F^{(m-1)} + a_{\ell m-1}F^{(m)}. \end{aligned}$$

This implies that $F^{(\ell+1)}$ is also a linear combination of $F, F^{(1)}, \dots, F^{(m-1)}$ over $\mathcal{M}$. Consequently, we see

$$r+1 := \operatorname{rank}(f_i^{(\ell)}; 0 \leq i,\ \ell \leq k) = \operatorname{rank}(f_i^{(\ell)}; 0 \leq i \leq k,\ \ell = 0, 1, 2, \dots).$$

Changing indices if necessary, we may assume that there are elements $a_0, \dots, a_r$ in $\mathcal{M}$ with $(a_0, \dots, a_r) \neq (0, \dots, 0)$ such that

$$a_0 f_0^{(\ell)} + \cdots + a_r f_r^{(\ell)} = 0$$

for all $\ell = 0, 1, \dots$. Take a point z_0 such that each a_i has no pole at z_0 and $(a_0(z_0), \dots, a_r(z_0)) \neq (0, \dots, 0)$ and consider the function

$$g = a_0(z_0)f_0 + \cdots + a_r(z_0)f_r.$$

Then $g^{(\ell)}(z_0) = a_0(z_0)f_0^{(\ell)}(z_0) + \cdots + a_r(z_0)f_r^{(\ell)}(z_0) = 0$ for all $\ell \geq 0$. By the theorem of identity, we see that g vanishes identically. Therefore, $f_0, \dots, f_r$ are linearly dependent over $\mathbf{C}$. This completes the proof of Proposition 2.1.7.

We now recall some properties of the exterior product $\bigwedge^{k+1}\mathbf{C}^{n+1}$.

DEFINITION 2.1.8. A vector A in $\bigwedge^{k+1}\mathbf{C}^{n+1}$ is called a *decomposable k-vector* if it is written as $A = A_0 \wedge \ldots \wedge A_k$ with $k+1$ vectors A_i $(0 \le i \le k)$ in $\mathbf{C}^{n+1}$.

Take a basis $\{E_0, \ldots, E_n\}$ of $\mathbf{C}^{n+1}$. Then, the set

$$\{E_{i_0} \wedge \ldots \wedge E_{i_k}; 0 \le i_0 < \cdots < i_k \le n\}$$

gives a basis of $\bigwedge^{k+1}\mathbf{C}^{n+1}$ and the space $\bigwedge^{k+1}\mathbf{C}^{n+1}$ is isomorphic with $\mathbf{C}^{N_k+1}$, where $N_k = \binom{n+1}{k+1} - 1$. We see easily that

(2.1.9) (i) *For* $k+1$ *vectors* $A_i = a_{i0}E_0 + \cdots + a_{in}E_n$ $(0 \le i \le k)$,

$$A_0 \wedge \ldots \wedge A_k = \sum_{0 \le j_0 < \cdots < j_k \le n} \det(a_{ij_\ell}; 0 \le i, \ell \le k) E_{j_0} \wedge \ldots \wedge E_{j_k}.$$

(ii) $A_0 \wedge \ldots \wedge A_k \ne 0$ *if and only if* $A_0, \ldots, A_k$ *are linearly independent.*

By $\{\{A_0, \ldots, A_k\}\}$ we denote the vector subspace of $\mathbf{C}^{n+1}$ generated by vectors $A_0, \ldots, A_k$.

(2.1.10) *If* $A_0, \ldots, A_k \in \mathbf{C}^{n+1}$ *are linearly independent, then we have*

$$\{\{A_0, \ldots, A_k\}\} = \{X \in \mathbf{C}^{n+1}; A_0 \wedge \ldots \wedge A_k \wedge X = 0\}.$$

To see this, we choose $n-k$ vectors $A_{k+1}, \ldots, A_n$ such that $A_0, \ldots, A_n$ give a basis of $\mathbf{C}^{n+1}$. Then, a vector $X = \sum_{i=0}^{n} c_i A_i$ is contained in $\{\{A_0, \ldots, A_k\}\}$ if and only if $c_{k+1} = \cdots = c_n = 0$. This is also equivalent to the condition

$$A_0 \wedge \ldots \wedge A_k \wedge X \equiv \sum_{i=k+1}^{n} c_i A_0 \wedge A_1 \wedge \ldots \wedge A_k \wedge A_i = 0,$$

because $A_0 \wedge A_1 \wedge \ldots \wedge A_k \wedge A_i (k+1 \le i \le n)$ are linearly independent. This shows (2.1.10).

PROPOSITION 2.1.11. *Let $A_0, \dots, A_k$ be linearly independent vectors in $\mathbf{C}^{n+1}$. For arbitrary vectors $B_0, \dots, B_k$, the following conditions are equivalent:*

(i) $\{\{A_0, \dots, A_k\}\} = \{\{B_0, \dots, B_k\}\}$,

(ii) $A_0 \wedge \dots \wedge A_k = cB_0 \wedge \dots \wedge B_k$ *for some* $c \neq 0$.

PROOF. Assume that (i) holds. Then, $B_0, \dots, B_k$ give a basis of both sides of (i). Each A_i may be written as $A_i = \sum_{i=0}^{k} c_{ij} B_j$ for some $c_{ij} \in \mathbf{C}$. Therefore, we have (ii) for $c := \det(c_{ij}) \neq 0$. On the other hand, by (2.1.10) we can easily conclude (i) from (ii).

We now consider the set $G(n,k)$ of all $(k+1)$-dimensional vector subspaces of $\mathbf{C}^{n+1}$ for each k $(0 \leq k \leq n)$. For each element $P^{N_k}(\mathbf{C})$ take a basis $\{A_0, \dots, A_k\}$ of P and consider the exterior product

$$A := A_0 \wedge \dots \wedge A_k \in \bigwedge^{k+1} \mathbf{C}^{n+1}.$$

Identifying the space $\bigwedge^{k+1} \mathbf{C}^{n+1}$ with $\mathbf{C}^{N_k+1}$ and letting π_k denote the canonical projection of $\mathbf{C}^{N_k+1} - \{0\}$ onto $\mathrm{P}^{N_k}(\mathbf{C})$, we set $\Phi(P) = \pi_k(A)$. By Proposition 2.1.11 $\Phi(P)$ is uniquely determined by P and the map Φ of $G(n,k)$ into $P^{N_k}(\mathbf{C})$ is injective, which is the so-called Plücker imbedding of $G(n,k)$. In this way, we regard $G(n,k)$ as a subset of $P^{N_k}(\mathbf{C})$. As is well-known, $G(n,k)$ may be considered as a compact submanifold of $P^{N_k}(\mathbf{C})$, which is called Grassmann manifold(cf., [43]).

Let f be a holomorphic curve in $P^n(\mathbf{C})$ with a reduced representation $f = (f_0 : \cdots : f_n)$. Take a holomorphic local coordinate z on an open set U. Set

$$F^{(\ell)} = (f_0^{(\ell)}, f_1^{(\ell)}, \dots, f_n^{(\ell)}) \qquad (\ell = 0, 1, \dots)$$

on U and define

$$F_k(z) = (F^{(0)} \wedge F^{(1)} \wedge \dots \wedge F^{(k)})(z) : U \to \bigwedge^{k+1} \mathbf{C}^{n+1}$$

for each $k = 0, 1, \dots, n$, where $F^{(0)} = F = (f_0, \dots, f_n)$. We see by (2.1.9), (i) that

$$F_k = \sum_{0 \leq i_0 < \cdots < i_k \leq n} W(f_{i_0}, \dots, f_{i_k}) E_{i_0} \wedge \dots \wedge E_{i_k}, \qquad (2.1.12)$$

where $E_i = (0, \dots, \overset{i\text{-th}}{1}, \dots, 0)$.

DEFINITION 2.1.13. A holomorphic map f of M into $P^n(\mathbf{C})$ is said to be *nondegenerate* if the image of f is not included in any hyperplane in $P^n(\mathbf{C})$.

If $f = (f_0 : \cdots : f_n)$ is nondegenerate, then $f_0, \ldots, f_n$ are linearly independent over $\mathbf{C}$. Therefore, by (2.1.12) and Proposition 2.1.7, $F_k \not\equiv 0$ for $0 \leq k \leq n-1$.

DEFINITION 2.1.14. Assume that f is nondegenerate. For $0 \leq k \leq n-1$ we define the *k-th derived curve* $f^k : M \to P^{N_k}(\mathbf{C})$ of f by

$$f^k = \pi_k \circ F_k,$$

namely, the holomorphic curve defined by the functions $W(f_{i_0}, \ldots, f_{i_k})$ $(0 \leq i_0 < \cdots < i_n \leq n)$, which is well-defined on the totality of M by the following:

REMARK 2.1.15. Let $f = (g_0 : g_1 : \ldots : g_n)$ be another reduced representation of f. Then there is a nowhere zero holomorphic function h such that $g_i = hf_i$. For vectors $G^{(\ell)} = (g_0^{(\ell)}, \ldots, g_n^{(\ell)})$ we have

$$G^{(\ell)} = hF^{(\ell)} + \binom{\ell}{1} h'F^{(\ell-1)} + \binom{\ell}{2} h''F^{(\ell-2)} + \cdots + h^{(\ell)}F$$

by the Leibniz formula. Therefore,

$$\begin{aligned} &G \wedge G' \wedge \ldots \wedge G^{(k)} \\ &\quad = hF \wedge (hF' + h'F) \wedge \ldots \wedge (hF^{(k)} + kh'G^{(k-1)} + \cdots + h^{(k)}F) \\ &\quad = h^{k+1}(F \wedge F' \wedge \ldots \wedge F^{(k)}). \end{aligned}$$

This shows that f^k does not depend on the choice of a reduced representation. Next take two holomorphic local coordinates ζ and z on an open set U. By $(F_k)_\zeta$ and $(F_k)_z$ we denote the functions defined as above with respect to the coordinates ζ and z respectively. By (2.1.12) and Proposition 2.1.6, we have

$$(F_k)_\zeta = (F_k)_z \left(\frac{dz}{d\zeta}\right)^{k(k+1)/2}.$$

This concludes that f^k does not depend on the choice of a holomorphic local coordinate. Therefore, f^k is well-defined on the totality of M.

The 0-th derived curve of f is nothing but the original map f itself.

§2.2 Frenet frames

We denote the hermitian product of two vectors $A = (a_0, \dots, a_n)$ and $B = (b_0, \dots, b_n)$ in $\mathbf{C}^{n+1}$ by

$$\langle A, B\rangle := (A, \bar{B}) = a_0\bar{b}_0 + \cdots + a_n\bar{b}_n.$$

Choose an orthonormal basis $\{E_0, \dots, E_n\}$ of $\mathbf{C}^{n+1}$. For two vectors

$$A = \sum_{0\le j_0<\cdots<j_k\le n} A_{j_0\dots j_k} E_{j_0} \wedge \dots \wedge E_{j_k}$$

and

$$B = \sum_{0\le j_0<\cdots<j_k\le n} B_{j_0\dots j_k} E_{j_0} \wedge \dots \wedge E_{j_k}$$

in $\bigwedge^{k+1} \mathbf{C}^{n+1}$ we define the hermitian product of A and B by

$$\langle A, B\rangle = \sum_{0\le j_0<\cdots<j_k\le n} A_{j_0\dots j_k} \bar{B}_{j_0\dots j_k}$$

and the norm of A by $|A| = \langle A, A\rangle^{1/2}$.

For decomposable vectors $A = A_0 \wedge \dots \wedge A_k$ and $B = B_0 \wedge \dots \wedge B_k$ we have

$$\langle A, B\rangle = \det(\langle A_i, B_j\rangle; 0 \le i, j \le k)$$

by (2.1.9) and the well-known formula for the determinant of the product of two matrices.

We now study a nondegenerate holomorphic map f of an open Riemann surface M into $P^n(\mathbf{C})$. Choosing a reduced representation $f = (f_0 : \cdots : f_n)$, we consider the maps $F_k : U \to \bigwedge^{k+1} \mathbf{C}^{n+1}$ defined in the previous section and their norms

$$|F_k| = \left(\sum_{0\le j_0<\cdots<j_k\le n} |W(f_{j_0}, \dots, f_{j_k})|^2 \right)^{1/2} \tag{2.2.1}$$

for each $k = 0, 1, \dots, n$.

We can prove the following:

PROPOSITION 2.2.2. *Assume that F_n has no zero on an open set U. For arbitrarily given C^∞ real-valued functions $t_k(z) (0 \leq k \leq n)$ on U there exists one and only one system $\{E_0, E_1, \dots, E_n\}$ of vector-valued C^∞ functions on U such that, for each point $z \in U$, $E_0(z), E_1(z), \dots, E_n(z)$ give an orthonormal basis of $\mathbf{C}^{n+1}$ and satisfy the condition*

$$(2.2.3)_k \qquad e^{it_k(z)} \frac{F_k(z)}{|F_k(z)|} = E_0(z) \wedge E_1(z) \wedge \dots \wedge E_k(z) \qquad (0 \leq k \leq n).$$

PROOF. We construct functions E_k satisfying $(2.2.3)_k$ by induction on k. For the case $k = 0$ the condition $(2.2.3)_0$ can be rewritten as $E_0 = e^{it_0(z)} F_0/|F_0|$ and so the desired function E_0 exists uniquely. Assume that there exist unique functions $E_0, \dots, E_k$ satisfying $(2.2.3)_\ell$ for $0 \leq \ell \leq k$. Furthermore, assume that a function E_{k+1} satisfies the condition $(2.2.3)_{k+1}$. Then $F^{(k+1)}(z) \in \{\{E_0(z), \dots, E_{k+1}(z)\}\}$ for each $z \in U$, and therefore we can write

$$F^{(k+1)} = c_0 E_0 + \dots + c_{k+1} E_{k+1}$$

for suitable functions c_i on U. Since $E_0(z), \dots, E_{k+1}(z)$ give an orthonormal basis, we have necessarily

$$(2.2.4) \qquad c_\ell = \langle F^{(k+1)}, E_\ell \rangle \qquad (0 \leq \ell \leq k).$$

We have also

$$\begin{aligned} F_{k+1} &= F_k \wedge F^{(k+1)} \\ &= e^{-it_k(z)} |F_k| E_0 \wedge \dots \wedge E_k \wedge F^{(k+1)} \\ &= e^{-it_k(z)} |F_k| E_0 \wedge \dots \wedge E_k \wedge (c_0 E_0 + \dots + c_{k+1} E_{k+1}) \\ &= c_{k+1} e^{-it_k(z)} |F_k| E_0 \wedge \dots \wedge E_k \wedge E_{k+1}. \end{aligned}$$

On the other hand, since

$$F_{k+1} = e^{-it_{k+1}(z)} |F_{k+1}| E_0 \wedge \dots \wedge E_{k+1}$$

by $(2.2.3)_{k+1}$, we see

$$c_{k+1} e^{-it_k(z)} |F_k| = e^{-it_{k+1}(z)} |F_{k+1}|.$$

Therefore,

$$c_{k+1} = e^{i(t_k - t_{k+1})} \frac{|F_{k+1}|}{|F_k|}. \tag{2.2.5}$$

This shows that all c_ℓ $(0 \leq \ell \leq k+1)$ are uniquely determined and that,

$$E_{k+1} = \frac{1}{c_{k+1}} (F^{(k+1)} - (c_0 E_0 + \ldots + c_k E_k)) \tag{2.2.6}$$

is also uniquely determined. They are obviously of class C^∞. Conversely, if we define the functions c_ℓ by (2.2.4) $\sim$ (2.2.5) and define E_{k+1} by (2.2.6), it is easily seen that $E_0, \ldots, E_k, E_{k+1}$ are mutually orthogonal. They satisfy also

$$\begin{aligned} E_0 \wedge \ldots \wedge E_k \wedge E_{k+1} &= \frac{1}{c_{k+1}} E_0 \wedge \ldots \wedge E_k \wedge F^{(k+1)} \\ &= \frac{e^{it_k}}{c_{k+1}} \frac{F_{k+1}}{|F_k|} = \frac{e^{it_{k+1}} F_{k+1}}{|F_{k+1}|}. \end{aligned}$$

Moreover, since

$$|E_{k+1}|^2 = |E_0 \wedge \ldots \wedge E_k \wedge E_{k+1}|^2 = |F_{k+1}|^2 / |F_{k+1}|^2 = 1,$$

E_{k+1} is a unit vector. The proof of Proposition 2.2.2 is completed.

The functions $E_0, E_1, \ldots, E_n$ given in Proposition 2.2.2 are an analogue of Frenet frames for a (real) curve in $\mathbf{R}^m$. We call them the *Frenet frames* of f on U with respect to the functions t_k.

REMARK 2.2.7. The assertion (2.2.5) may be rewritten as

$$\langle F^{(k+1)}, E_{k+1} \rangle = e^{i(t_k - t_{k+1})} \frac{|F_{k+1}|}{|F_k|},$$

which is useful in the following discussions.

For a system of Frenet frames $E_0(z), \ldots, E_n(z)$ we set

$$dE_k = \sum_{\ell=0}^{n} \theta_{k\ell} E_\ell \qquad (0 \leq k \leq n), \tag{2.2.8}$$

where $\theta_{k\ell}$ are forms of degree one. They have the following properties.

PROPOSITION 2.2.9. (i) $\theta_{k\ell} + \bar{\theta}_{\ell k} = 0$.

(ii) $d\theta_{k\ell} = \sum_{m=0}^{n} \theta_{km} \wedge \theta_{m\ell}$.

(iii) *If* $|k - \ell| > 1$, *then* $\theta_{k\ell} = 0$ *and therefore we may write*

$$dE_k = \theta_{kk-1} E_{k-1} + \theta_{kk} E_k + \theta_{kk+1} E_{k+1},$$

where θ_{kk+1} *is a* $(1,0)$*-form and* θ_{kk-1} *is a* $(0,1)$*-form.*

(iv) *For each* ζ *we can find a system of Frenet frames around* ζ *satisfying the condition that*

$$\theta_{kk}(\zeta) = 0 \qquad (0 \le k \le n).$$

PROOF. (i) By definition,

$$\langle E_k, E_\ell \rangle = \delta_{k\ell} = \begin{cases} 1 & k = \ell \\ 0 & k \neq \ell \end{cases}$$

for each k, ℓ. Differentiating both sides of this, we have

$$\langle dE_k, E_\ell \rangle + \langle E_k, dE_\ell \rangle = 0.$$

By substituting (2.2.8) into this identity, we easily obtain the desired identity.

(ii) Differentiate both sides of (2.2.8). We then have

$$\begin{aligned} 0 &= \sum_{\ell=0}^{n} (d\theta_{k\ell} E_\ell - \theta_{k\ell} \wedge dE_\ell) \\ &= \sum_{\ell=0}^{n} d\theta_{k\ell} E_\ell - \sum_{\ell=0}^{n} \sum_{m=0}^{n} \theta_{k\ell} \wedge \theta_{\ell m} E_m \\ &= \sum_{\ell=0}^{n} \left(d\theta_{k\ell} - \sum_{m=0}^{n} \theta_{km} \wedge \theta_{m\ell} \right) E_\ell. \end{aligned}$$

This gives the assertion (ii).

(iii) Set

$$E_k = a_0 F + a_1 F' + \cdots + a_k F^{(k)},$$

where a_ℓ are functions on a neighborhood of a point ζ. Then, since $(\partial/\partial\bar{z}) F^{(\ell)} = 0$ and $(\partial/\partial z) F^{(\ell)} = F^{(\ell+1)}$ for all ℓ,

$$\begin{aligned} dE_k = &\left(\sum_{\ell=0}^{k} \frac{\partial a_\ell}{\partial \bar{z}} F^{(\ell)} \right) d\bar{z} + \left(\sum_{\ell=0}^{k} \frac{\partial a_\ell}{\partial z} F^{(\ell)} \right) dz \\ &+ \left(\sum_{\ell=0}^{k-1} a_\ell F^{(\ell+1)} \right) dz + a_k F^{(k+1)} dz. \end{aligned} \tag{2.2.10}$$

All but the last terms of the right hand side are written as linear combinations of $E_0, \dots, E_k$ and $F^{(k+1)}$ is a linear combination of $E_0, \dots, E_{k+1}$, because

$$\{\{F, F', \dots, F^{(\ell)}\}\} = \{\{E_0, \dots, E_\ell\}\} \qquad (\ell = k, k+1).$$

This implies that

$$\begin{aligned} \theta_{k\ell} &= \langle dE_k, E_\ell \rangle = 0 & \text{if} \quad \ell > k+1, \\ \theta_{k\ell} &= -\bar{\theta}_{\ell k} = 0 & \text{if} \quad k > \ell + 1. \end{aligned}$$

Moreover, by observing the (0, 1)-components of both sides of (2.2.10) we can conclude that $\bar{\partial} E_k$ is a linear combination of $E_0, \dots, E_k$, whence θ_{kk+1} is of type (1, 0) and $\theta_{kk-1} = -\bar{\theta}_{k-1k}$ is of type (0, 1). This implies that the assertion (iii) is valid.

(iv) Take a system of Frenet frames constructed as in Proposition 2.2.2. For arbitrarily given real-valued functions s_k we consider the frames $\tilde{E}_k = e^{is_k} E_k$. By (iii)

$$d\tilde{E}_k = e^{is_k} \theta_{kk-1} E_{k-1} + e^{is_k}(\theta_{kk} + \sqrt{-1} ds_k) E_k + e^{is_k} \theta_{kk+1} E_{k+1}.$$

If we choose a real-valued function s_k such that $\theta_{kk} = -\sqrt{-1} ds_k$ at ζ, the functions $\tilde{E}_k$ satisfy the desired condition. This is possible because $\sqrt{-1}\theta_{kk}$ is real. The proof of Proposition 2.2.9 is completed.

The (normalized) Fubini-Study metric form on $P^{N_k}(\mathbf{C})$ is defined to be the form ω_k such that

$$\pi_k^*(\omega_k) = dd^c \log(|W_0|^2 + \cdots + |W_{N_k}|^2),$$

where $d^c = (\sqrt{-1}/4\pi)(\bar{\partial} - \partial)$, $\pi_k : \mathbf{C}^{N_k+1} - \{0\} \to P^{N_k}(\mathbf{C})$ is the canonical projection and $W_0, \dots, W_{N_k}$ are the standard coordinates on $\mathbf{C}^{N_k+1}$. We denote by Ω_k the pull-back of ω_k via the derived curve f^k. Then

$$\Omega_k = dd^c \log |F_k|^2$$

on $\{z; F_k(z) \neq 0\}$. For a point ζ with $F_k(\zeta) = 0$, by taking a holomorphic local coordinate z on a neighborhood U of ζ, we can write $|F_k|$ as is (2.2.1). Therefore, we can find a positive differentiable function v on a neighborhood U of ζ such that

$$|F_k(z)|^2 = |z - \zeta|^{2m_k} v(z),$$

where $m_k = \nu_{|F_k|}(\zeta)$. We then have $\Omega_k = dd^c \log v(z)$ on $U - \{\zeta\}$, which is extended to a differential form on U.

PROPOSITION 2.2.11. *For a holomorphic local coordinate z, set $\Omega_k = h_k dd^c|z|^2$ for $0 \leq k \leq n-1$. Then,*

$$h_k = \frac{|F_{k-1}|^2 |F_{k+1}|^2}{|F_k|^4}$$

on the set $\{\zeta; F_k(\zeta) \neq 0\}$, where we set $|F_{-1}| \equiv 1$ for the sake of convenience.

PROOF. Take a point ζ with $F_k(\zeta) \neq 0$. By definition we see easily that

$$\begin{aligned} h_k &= \frac{\partial^2}{\partial z \partial \bar{z}} \log \langle F \wedge \ldots \wedge F^{(k)}, F \wedge \ldots \wedge F^{(k)} \rangle \\ &= \frac{\partial}{\partial z} \left(\frac{\langle F \wedge \ldots \wedge F^{(k)}, F \wedge \ldots \wedge F^{(k-1)} \wedge F^{(k+1)} \rangle}{|F_k|^2} \right) \\ &= \frac{|F \wedge \ldots \wedge F^{(k-1)} \wedge F^{(k+1)}|^2}{|F_k|^2} \\ &\quad - \frac{|\langle F \wedge \ldots \wedge F^{(k)}, F \wedge \ldots \wedge F^{(k-1)} \wedge F^{(k+1)} \rangle|^2}{|F_k|^4}. \end{aligned}$$

According to Proposition 2.2.2 we may take the Frenet frame $\{E_0, E_1, \ldots, E_n\}$ with respect to the functions $t_\ell(\zeta) \equiv 0 (0 \leq \ell \leq n)$ and write

$$F^{(k+1)} = c_0 E_0 + \cdots + c_k E_k + c_{k+1} E_{k+1}$$

with suitable functions c_ℓ. Then, by $(2.2.3)_{k-1}$ and the fact that $E_0 \wedge \ldots \wedge E_{k-1} \wedge E_k$ and $E_0 \wedge \ldots \wedge E_{k-1} \wedge E_{k+1}$ are mutually orthogonal, we have

$$\begin{aligned} &|F \wedge \ldots \wedge F^{(k-1)} \wedge F^{(k+1)}|^2 \\ &\quad = |F_{k-1}|^2 |E_0 \wedge \ldots \wedge E_{k-1} \wedge (c_k E_k + c_{k+1} E_{k+1})|^2 \\ &\quad = |F_{k-1}|^2 (|c_k|^2 + |c_{k+1}|^2). \end{aligned}$$

Similarly, we get

$$\begin{aligned} &|\langle F \wedge \ldots \wedge F^{(k)}, F \wedge \ldots \wedge F^{(k-1)} \wedge F^{(k+1)} \rangle|^2 \\ &\quad = |F_k|^2 |F_{k-1}|^2 |c_k|^2. \end{aligned}$$

Therefore, we conclude

$$h_k = \frac{|F_k|^2|F_{k-1}|^2((|c_k|^2 + |c_{k+1}|^2) - |c_k|^2)}{|F_k|^4}$$
$$= \frac{|F_{k-1}|^2|c_{k+1}|^2}{|F_k|^2}.$$

On the other hand, by Remark 2.2.7 we see that

$$|c_{k+1}|^2 = \frac{|F_{k+1}|^2}{|F_k|^2},$$

whence we can easily conclude Proposition 2.2.11.

We now take Frenet frames $E_0, \dots, E_n$ with respect to the functions $t_\ell(z) \equiv 0$ and consider 1-forms $\theta_{k\ell}$ satisfying the identities (2.2.8). Then we have the following:

PROPOSITION 2.2.12. $\Omega_k = \dfrac{\sqrt{-1}}{2\pi}\theta_{kk+1} \wedge \bar{\theta}_{kk+1} \qquad (0 \leq k \leq n-1).$

PROOF. As in the proof of Proposition 2.2.9, we set

$$E_k = a_0 F + a_1 F' + \cdots + a_k F^{(k)}.$$

Then, according to (2.2.10) and Remark 2.2.7,

$$\theta_{kk+1} = \langle dE_k, E_{k+1}\rangle$$
$$= a_k \langle F^{(k+1)}, E_{k+1}\rangle dz$$
$$= a_k \frac{|F_{k+1}|}{|F_k|} dz.$$

On the other hand, since $\langle F^{(\ell)}, E_k\rangle = 0$ for $\ell = 0, 1, \dots, k-1$,

$$1 = \langle E_k, E_k\rangle = \langle a_0 F + a_1 F' + \ldots + a_k F^{(k)}, E_k\rangle$$
$$= a_k \langle F^{(k)}, E_k\rangle$$
$$= a_k \frac{|F_k|}{|F_{k-1}|}$$

using Remark 2.2.7. It follows that

$$\theta_{kk+1} = \frac{|F_{k-1}|}{|F_k|}\frac{|F_{k+1}|}{|F_k|} dz$$
$$= \frac{|F_{k-1}||F_{k+1}|}{|F_k|^2} dz.$$

This implies the desired identity by virtue of Proposition 2.2.11.

§2.3 Contact functions

We now recall the basic properties of the interior product of vectors in the exterior algebra.

Take two vectors $A \in \bigwedge^{k+1} \mathbf{C}^{n+1}$ and $B \in \bigwedge^{h+1} \mathbf{C}^{n+1}$, where $k \geq h$. As is easily seen, there is one and only one vector $C \in \bigwedge^{k-h} \mathbf{C}^{n+1}$ satisfying the condition

$$\langle C, D\rangle = \langle A, B \wedge D\rangle \qquad \text{for all} \quad D \in \bigwedge^{k-h} \mathbf{C}^{n+1}.$$

DEFINITION 2.3.1. The vector C satisfying the above condition is called the *interior product* of vectors A and B, which we denote by $A \vee B$ after [68].

For the particular case $k = h$, $A \vee B = \langle A, B\rangle$.

For later use, we give a more precise expression of the interior product of a decomposable k-vector $A = A_0 \wedge \ldots \wedge A_k$ and a vector B in $\mathbf{C}^{n+1}$. Taking an arbitrary orthonormal basis $\{E_0, \ldots, E_n\}$ of $\mathbf{C}^{n+1}$, we write

$$B = b_0 E_0 + b_1 E_1 + \cdots + b_n E_n$$

and

$$A_i = a_{i0} E_0 + a_{i1} E_1 + \cdots + a_{in} E_n \qquad (0 \leq i \leq k).$$

Set $a_{j_0 \ldots j_k} := \det(a_{ij_\ell}; 0 \leq i, \ell \leq k)$ for $0 \leq j_0, \ldots, j_k \leq n$. By (2.1.9) the vector A may be written as

$$A = \sum_{0 \leq j_0 < \ldots < j_k \leq n} a_{j_0 \ldots j_k} E_{j_0} \wedge \ldots \wedge E_{j_k}.$$

Write

$$A \vee B = \sum_{0 \leq j_1 < \cdots < j_k \leq n} c_{j_1 \cdots j_k} E_{j_1} \wedge \ldots \wedge E_{j_k} \qquad (c_{j_1 \ldots j_k} \in \mathbf{C}).$$

We then have

$$\begin{aligned} c_{j_1 \ldots j_k} &= \langle A \vee B, E_{j_1} \wedge \ldots \wedge E_{j_k}\rangle \\ &= \langle A, B \wedge E_{j_1} \wedge \ldots \wedge E_{j_k}\rangle \\ &= \sum_{j=0}^{n} \bar{b}_j \langle A, E_j \wedge E_{j_1} \wedge \ldots \wedge E_{j_k}\rangle \\ &= \sum_{j \neq j_1, \cdots, j_k} a_{j j_1 \cdots j_k} \bar{b}_j. \end{aligned}$$

Therefore, we see

(2.3.2) *For* $A = A_0 \wedge \ldots \wedge A_k$ *with* $A_i = \sum_{j=0}^{n} a_{ij} E_j$ *and* $B = \sum_{j=0}^{n} b_j E_j$,

$$A \vee B = \sum_{0 \le j_1 < \ldots < j_k \le n} \left(\sum_{j_0 \neq j_1, \ldots, j_k} \det(a_{ij_\ell}; 0 \le i, \ell \le k) \bar{b}_{j_0} \right) E_{j_1} \wedge \ldots \wedge E_{j_k}.$$

Assume that $A \neq 0$, or equivalently, $A_0, \ldots, A_k$ are linearly independent. We choose an orthonormal basis $\{E_0, \ldots, E_n\}$ of $\mathbf{C}^{n+1}$ such that $\{E_0, \ldots, E_k\}$ gives a basis of $\{\{A_0, \ldots, A_k\}\}$. Then, we can write simply

$$A = a_{0\ldots k} E_0 \wedge \ldots \wedge E_k.$$

Substitute $a_{ij} = \delta_{ij}$ for $i = 0, \ldots, k$ and $j = 0, \ldots, n$ into the identity of (2.3.2). We have easily

$$A \vee B = a_{0\ldots k} \sum_{j=0}^{k} (-1)^j \bar{b}_j E_0 \wedge \ldots \wedge E_{j-1} \wedge E_{j+1} \wedge \ldots \wedge E_k.$$

This implies that

$$(2.3.3) \qquad |A \vee B|^2 = |A|^2 \left(\sum_{j=0}^{k} |b_j|^2 \right),$$

because $|A| = |a_{0\ldots k}|$ in this case. Therefore, we always have

$$(2.3.4) \qquad |A \vee B| \le |A||B|$$

and, we can conclude that

(2.3.5) $A \vee B = 0$ *if and only if* $b_0 = \cdots = b_k = 0$, *which means that* B *is orthogonal to the space* $\{\{A_0, \ldots, A_k\}\}$.

Now, we consider a nondegenerate holomorphic map f of an open Riemann surface M into $P^n(\mathbf{C})$ with a reduced representation $f = (f_0 : \cdots : f_n)$ and a hyperplane

$$H: \langle W, A \rangle \equiv \bar{a}_0 w_0 + \cdots + \bar{a}_n w_n = 0$$

for a nonzero vector $A = (a_0, \ldots, a_n)$ in $\mathbf{C}^{n+1}$, where we may assume $|A| = 1$. For each $p \in M$, taking a holomorphic local coordinate z on a neighborhood of p, we define the functions F_k as in the previous sections.

DEFINITION 2.3.6. Set $F_k(H) = F_k \vee A$. We define *the k-th contact function* of f for H (or for A) by

$$\varphi_k(H)(z) = \frac{|F_k(H)(z)|^2}{|F_k(z)|^2}.$$

REMARK 2.3.7. (i) Set $F(H) := \bar{a}_0 f_0 + \cdots + \bar{a}_n f_n$. Since $F_0 \vee A = F(H)$, one has

$$\varphi_0(H) = \frac{|F(H)|^2}{|F|^2}.$$

On the other hand,

$$\varphi_n(H) = \frac{|W(f_0, \ldots, f_n)|^2 |A|^2}{|F_n|^2} = 1.$$

(ii) For each point ζ the value $\lim_{z \to \zeta} \varphi_k(H)(z)$ exists. In fact, we can write

$$|F_k(H)(z)|^2 = |z - \zeta|^\ell h_1(z), \quad |F_k(z)|^2 = |z - \zeta|^m h_2(z)$$

with positive $\mathbf{C}^\infty$ functions $h_i(z)$ $(i = 1, 2)$ and nonnegative integers ℓ and m. Here, $\ell \geq m$ by (2.3.4). This guarantees the existence of the above limit.

(iii) The contact functions do not depend on a choice of a reduced representation of f or of a holomorphic local coordinate z.

For, we can write

$$\text{(2.3.8)} \quad \begin{aligned} |F_k(H)|^2 &= \sum_{0 \leq i_1 < \cdots < i_k \leq n} \left| \sum_{i_0 \neq i_1, \cdots, i_k} \bar{a}_{i_0} W(f_{i_0}, \ldots, f_{i_k}) \right|^2, \\ |F_k|^2 &= \sum_{0 \leq i_0 < \cdots < i_k \leq n} |W(f_{i_0}, \ldots, f_{i_k})|^2. \end{aligned}$$

Both of these are multiplied by the same factor $\left| \frac{dz}{d\zeta} \right|^{k(k+1)}$ by Proposition 2.1.6 if we choose another holomorphic local coordinate ζ and they are multiplied by $|h|^{2(k+1)}$ if we choose another reduced representation $f = (hf_0 : \cdots : hf_n)$ given by a nowhere zero holomorphic function h.

(iv) The contact functions $\varphi_k(H)$ do not vanish identically for $k = 0, \ldots, n$.

For, by (2.3.8), $F_k(H)(z) \equiv 0$ when and only when

$$\sum_{j \neq i_1, \dots, i_k} \bar{a}_j W(f_j, f_{i_1}, \dots, f_{i_k}) \equiv 0$$

for all $i_1, \dots, i_k$. If $F_k(H) \equiv 0$ for some k, then

$$W(F(H), f_{i_1}, \dots, f_{i_k}) \equiv 0$$

for all $i_1, \dots, i_k$. By Proposition 2.1.7 this implies that $F(H), f_{i_1}, \dots, f_{i_k}$ are linearly dependent, which contradicts the nondegeneracy of f.

For brevity, we set $\varphi_k = \varphi_k(H)(0 \leq k \leq n-1)$ in the following.

PROPOSITION 2.3.9. $d\varphi_k \wedge d^c\varphi_k = (\varphi_{k+1} - \varphi_k)(\varphi_k - \varphi_{k-1})\Omega_k$.

For the proof of this, it suffices to show that the given identity holds at each point ζ with $F_n(\zeta) \neq 0$ because both sides are continuous. We take a system of Frenet frames $E_0, \dots, E_n$ of f around ζ satisfying the condition of Proposition 2.2.2. Then the vector A can be represented as

$$A = a_0 E_0 + \cdots + a_n E_n$$

for suitable functions $a_j(z)$ around ζ.

We first show the following;

(2.3.10) *For the above contact functions it holds that*

(i) $\varphi_k = |a_0|^2 + \cdots + |a_k|^2$,

(ii) $\partial\varphi_k = \langle E_0 \wedge \ldots \wedge E_{k-1} \wedge E_{k+1} \vee A, E_0 \wedge \ldots \wedge E_k \vee A\rangle \theta_{kk+1} = \bar{a}_{k+1} a_k \theta_{kk+1}$.

The assertion (i) is easily seen using (2.3.3). In fact, by considering a k-vector $E_0 \wedge \ldots \wedge E_k$ and a vector A as A and B of (2.3.3) respectively, we have

$$|E_0 \wedge \ldots \wedge E_k \vee A|^2 = |a_0|^2 + \cdots + |a_k|^2.$$

To see (ii), we denote by $\theta_{kk}^{(1)}$ and $\theta_{kk}^{(2)}$ the (1, 0)-component and (0, 1)-component of θ_{kk} respectively. Since $\theta_{kk} = -\bar{\theta}_{kk}$, we see

$$\theta_{kk}^{(1)} + \bar{\theta}_{kk}^{(2)} = 0.$$

By observing the (0, 1)-component and (1, 0)-component of both sides of the identity of Proposition 2.2.9, (iii), we have

$$\partial E_k = \theta_{kk+1} E_{k+1} + \theta^{(1)}_{kk} E_k,$$
$$\bar\partial E_k = \theta_{kk-1} E_{k-1} + \theta^{(2)}_{kk} E_k.$$

We use the notation $\varphi_k(A) = E_0 \wedge \ldots \wedge E_k \vee A$. Substitute the above identities into the formula

$$\begin{aligned}\partial\varphi_k &= \partial\langle \varphi_k(A), \varphi_k(A)\rangle \\ &= \sum_{\ell=0}^{k} \langle E_0 \wedge \ldots \wedge E_{\ell-1} \wedge \partial E_\ell \wedge E_{\ell+1} \wedge \ldots \wedge E_k \vee A, \varphi_k(A)\rangle \\ &\quad + \langle \varphi_k(A), E_0 \wedge \ldots \wedge E_{\ell-1} \wedge \bar\partial E_\ell \wedge E_{\ell+1} \wedge \ldots \wedge E_k \vee A\rangle.\end{aligned}$$

We then have

$$\begin{aligned}\partial\varphi_k &= \left(\sum_{\ell=1}^{k} \theta^{(1)}_{\ell\ell}\right) \langle \varphi_k(A), \varphi_k(A)\rangle + \left(\sum_{\ell=1}^{k} \bar\theta^{(2)}_{\ell\ell}\right) \langle \varphi_k(A), \varphi_k(A)\rangle \\ &\quad + \theta_{kk+1} \langle E_0 \wedge \ldots \wedge E_{k-1} \wedge E_{k+1} \vee A, \varphi_k(A)\rangle \\ &= \theta_{kk+1} \langle E_0 \wedge \ldots \wedge E_{k-1} \wedge E_{k+1} \vee A, \varphi_k(A)\rangle.\end{aligned}$$

This shows the first identity of (ii). The last identity of (ii) is easily obtained by replacing A in the middle term by a linear combination of E_ℓ.

PROOF OF PROPOSITION 2.3.9. Proposition 2.3.9 is now obvious. In fact, it holds that

$$\begin{aligned}d\varphi_k \wedge d^c\varphi_k &= \frac{\sqrt{-1}}{2\pi} \partial\varphi_k \wedge \bar\partial\varphi_k \\ &= \bar a_{k+1} a_k a_{k+1} \bar a_k \frac{\sqrt{-1}}{2\pi} \theta_{kk+1} \wedge \bar\theta_{kk+1} \\ &= (\varphi_{k+1} - \varphi_k)(\varphi_k - \varphi_{k-1})\Omega_k\end{aligned}$$

by the use of Proposition 2.2.12 and (2.3.10).

PROPOSITION 2.3.11. *It holds that*

$$dd^c \log \varphi_k = \frac{\varphi_{k+1}\varphi_{k-1} - {\varphi_k}^2}{{\varphi_k}^2} \Omega_k.$$

PROOF. For the proof of Proposition 2.3.11, we may assume that the property of (iv) of Proposition 2.2.9 holds at ζ for suitably chosen real-valued functions $t_k(z)$.

By direct calculations, we see

$$dd^c \log \varphi_k = \frac{\sqrt{-1}}{2\pi} \frac{\varphi_k \partial\bar{\partial}\varphi_k - \partial\varphi_k \wedge \bar{\partial}\varphi_k}{{\varphi_k}^2}.$$

We study the first term of the numerator of the right hand side. Since

$$\bar{\partial}\varphi_k = \langle E_0 \wedge \ldots \wedge E_k \vee A, E_0 \wedge \ldots \wedge E_{k-1} \wedge E_{k+1} \vee A\rangle \bar{\theta}_{kk+1},$$

we have

$$\begin{aligned}\partial\bar{\partial}\varphi_k = {} & \partial(\langle E_0 \wedge \ldots \wedge E_k \vee A, E_0 \wedge \ldots \wedge E_{k-1} \wedge E_{k+1} \vee A\rangle)\bar{\theta}_{kk+1} \\ & + \langle E_0 \wedge \ldots \wedge E_k \vee A, E_0 \wedge \ldots \wedge E_{k-1} \wedge E_{k+1} \vee A\rangle \partial\bar{\theta}_{kk+1}.\end{aligned}$$

By Proposition 2.2.9, (iv), we see

$$d\theta_{kk+1} = \theta_{kk} \wedge \theta_{kk+1} + \theta_{kk+1} \wedge \theta_{k+1k+1} = 0$$

at ζ and so

$$\bar{\partial}\theta_{kk+1} = 0$$

there. Therefore,

$$\partial\bar{\partial}\varphi_k = \partial(\langle E_0 \wedge \ldots \wedge E_k \vee A, E_0 \wedge \ldots \wedge E_{k-1} \wedge E_{k+1} \vee A\rangle)\bar{\theta}_{kk+1}$$

at ζ. On the other hand, since

$$\partial E_\ell = \theta_{\ell\ell+1} E_{\ell+1}, \qquad \bar{\partial} E_\ell = \theta_{\ell\ell-1} E_{\ell-1}$$

at ζ by Proposition 2.2.9, (iii) and (iv), we get

$$\begin{aligned}\partial\bar{\partial}\varphi_k = {} & (\langle E_0 \wedge \ldots \wedge E_{k-1} \wedge E_{k+1} \vee A, E_0 \wedge \ldots \wedge E_{k-1} \wedge E_{k+1} \vee A\rangle \\ & - \langle E_0 \wedge \ldots \wedge E_k \vee A, E_0 \wedge \ldots \wedge E_k \vee A\rangle)\theta_{kk+1} \wedge \bar{\theta}_{kk+1}\end{aligned}$$

by the same calculations as in the proof of (2.3.10), (ii). Therefore, by (2.3.3) and (2.3.10), (i), we have

$$\begin{aligned}& \partial\bar{\partial}\varphi_k \\ & = (|a_0|^2 + \cdots + |a_{k-1}|^2 + |a_{k+1}|^2 - |a_0|^2 - \cdots - |a_k|^2)\theta_{kk+1} \wedge \bar{\theta}_{kk+1}, \\ & = (|a_{k+1}|^2 - |a_k|^2)\theta_{kk+1} \wedge \bar{\theta}_{kk+1}, \\ & = (\varphi_{k+1} - \varphi_k - (\varphi_k - \varphi_{k-1}))\theta_{kk+1} \wedge \bar{\theta}_{kk+1}.\end{aligned}$$

This implies that

$$\begin{aligned} dd^c \log \varphi_k &= \frac{\varphi_k(\varphi_{k+1} - 2\varphi_k + \varphi_{k-1}) - (\varphi_{k+1} - \varphi_k)(\varphi_k - \varphi_{k-1})}{\varphi_k^2}\Omega_k \\ &= \frac{\varphi_{k+1}\varphi_{k-1} - \varphi_k^2}{\varphi_k^2}\Omega_k. \end{aligned}$$

§2.4 Nochka weights for hyperplanes in subgeneral position

For later use, we shall give some linear algebraic properties which were given by E. I. Nochka in [57](cf, [9]).

Let $N \geq n$ and $q \geq N+1$. We consider q hyperplanes $H_j (1 \leq j \leq q)$ in $P^n(\mathbf{C})$, which are given by

$$H_j : \langle W, A_j \rangle = 0 \qquad (1 \leq j \leq q)$$

for nonzero vectors A_j in $\mathbf{C}^{n+1}$.

Following W. Chen([9]), we give the following:

DEFINITION 2.4.1. We say that $H_1, \ldots, H_q$ (or $A_1, \ldots, A_q$) are in N-*subgeneral* position if for every $1 \leq i_0 < \cdots < i_N \leq q$

$$\{\{A_{i_0}, A_{i_1}, \ldots, A_{i_N}\}\} = \mathbf{C}^{n+1}.$$

By definition, $\{H_j; 1 \leq j \leq q\}$ are in n-subgeneral position if and only if they are in general position.

We assume that $H_1, \ldots, H_q$ are in N-subgeneral position in the followings.

Set $Q := \{1, 2, \ldots, q\}$. For each subset R of Q we set

$$d(R) := \dim\{\{A_j; j \in R\}\}$$

and $d(\emptyset) := 0$ for convenience's sake. We set also $d_R(S) := d(S) - d(R)$ and

$$\varepsilon_R(S) := \frac{d_R(S)}{\#S - \#R}$$

for each R, S with $R \subset S \subseteq Q$. Obviously, $0 \leq \varepsilon_R(S) \leq 1$.

(2.4.2) *Let $R \subset S_i \subseteq Q$ for $i = 1, 2$. Then,*

$$d_R(S_1 \cup S_2) + d_R(S_1 \cap S_2) \le d_R(S_1) + d_R(S_2).$$

To see this, we set $V_i = \{\{A_j; j \in S_i\}\}$ $(i = 1, 2)$. We then have

$$\dim(V_1 + V_2) + \dim(V_1 \cap V_2) = \dim V_1 + \dim V_2.$$

Since $d(S_1 \cap S_2) \le \dim(V_1 \cap V_2)$ and $d(S_1 \cup S_2) = \dim(V_1 + V_2)$, we get

$$d(S_1 \cup S_2) + d(S_1 \cap S_2) \le d(S_1) + d(S_2).$$

By subtracting $2d(R)$ from both sides of this, we conclude (2.4.2).

(2.4.3) *Let $R_1 \subseteq R_2 \subseteq Q$. If $\# R_2 \le N + 1$, then*

$$\# R_1 - d(R_1) \le \# R_2 - d(R_2) \le N - n.$$

PROOF. By (2.4.2), we obtain

$$d(R_2) = d(R_1 \cup (R_2 - R_1)) \le d(R_1) + d(R_2 - R_1).$$

Since $d(R_2 - R_1) \le \#(R_2 - R_1) = \# R_2 - \# R_1$, we have the first inequality of (2.4.3). Choose R_3 with $R_2 \subseteq R_3$ and $\# R_3 = N + 1$. Then $d(R_3) = n + 1$ by the N-nondegeneracy of $\{H_j\}$ and so

$$\# R_2 - d(R_2) \le \# R_3 - d(R_3) = N - n.$$

PROPOSITION 2.4.4. *Let $H_1, \ldots, H_q$ be hyperplanes in N-subgeneral position in $P^n(\mathbf{C})$, where $N > n$ and $q \ge 2N - n + 1$. Then, there exists a family of subsets $\{N_i; 0 \le i \le s\}$ of $Q := \{1, 2, \ldots, q\}$ satisfying the following conditions:*

(i) $N_0 = \emptyset \subset N_1 \subset \cdots \subset N_s$ *and* $d(N_s) < n + 1$,

(ii) $\varepsilon_{N_0}(N_1) < \varepsilon_{N_1}(N_2) < \cdots < \varepsilon_{N_{s-1}}(N_s) < \dfrac{n + 1 - d(N_s)}{2N - n + 1 - \# N_s}$,

(iii) *for every R with $N_{i-1} \subset R \subseteq Q$ $(1 \le i \le s)$, if $d(N_{i-1}) < d(R) < n + 1$, then*

$$\varepsilon_{N_{i-1}}(N_i) \le \varepsilon_{N_{i-1}}(R)$$

and, moreover, $\varepsilon_{N_{i-1}}(N_i) = \varepsilon_{N_{i-1}}(R)$ only when $R \subseteq N_i$,

(iv) *for every R with $N_s \subset R \subseteq Q$, if $d(N_s) < d(R) < n+1$, then*

$$\varepsilon_{N_s}(R) \geq \frac{n+1-d(N_s)}{2N-n+1-\#N_s}.$$

PROOF. We start the proof by setting $N_0 := \emptyset$. It suffices to show that, under the assumption that there is a family $\{N_i; 0 \leq i \leq s\}$ satisfying conditions (i) $\sim$ (iii), it satisfies also condition (iv) or else we can construct a subset N_{s+1} such that the family $\{N_i; 0 \leq i \leq s+1\}$ satisfies conditions (i) $\sim$ (iii). In fact, if the latter case occurs, we can reach the desired conclusion after finitely many repetitions of these constructions.

Now, consider a family $\{N_i; 0 \leq i \leq s\}$ satisfying conditions (i) $\sim$ (iii). Set

$$\mathcal{R} := \{R; N_s \subseteq R \subset Q, d(N_s) < d(R) < n+1\}.$$

Then each $R \in \mathcal{R}$ satisfies $\#R \leq N$. For, otherwise, $d(R) \geq n+1$ by the assumption of N-nondegeneracy. If $\mathcal{R} = \emptyset$, N_s satisfies the condition (iv) and so the proof is accomplished, so that we may assume that $\mathcal{R} \neq \emptyset$. For brevity, set $d_s(R) := d_{N_s}(R), \varepsilon_s(R) := \varepsilon_{N_s}(R)$ and $n_s(R) := \#R - \#N_s$ for each $R \in \mathcal{R}$. Moreover, for convenience's sake, set $\varepsilon_{-1}(N_0) := 0$. Consider the value

$$\varepsilon_0 := \min\{\varepsilon_s(R); R \in \mathcal{R}\}.$$

By condition (iii), for all $R \in \mathcal{R}$, $\varepsilon_{s-1}(N_s) \leq \varepsilon_{s-1}(R)$ holds and moreover $\varepsilon_{s-1}(N_s) < \varepsilon_{s-1}(R)$ holds because $R \not\subseteq N_s$. Then, we easily see that $\varepsilon_{s-1}(R) < \varepsilon_s(R)$ and therefore

$$\varepsilon_{s-1}(N_s) < \varepsilon_0. \tag{2.4.5}$$

For our purpose, it may be assumed that

$$\varepsilon_0 < \frac{n+1-d(N_s)}{2N-n+1-\#N_s} \ (\leq 1). \tag{2.4.6}$$

because, otherwise, (iv) holds. Define

$$\tilde{\mathcal{R}} = \{R; R \in \mathcal{R}, \varepsilon_s(R) = \varepsilon_0\}.$$

We shall prove the following:

(2.4.7) *For every $R_1, R_2 \in \tilde{\mathcal{R}}$, the set $R_1 \cup R_2$ is also contained in $\tilde{\mathcal{R}}$.*

PROOF OF (2.4.7). By definition, we have

$$\varepsilon_0 = \frac{d_s(R_1)}{n_s(R_1)} = \frac{d_s(R_2)}{n_s(R_2)}. \tag{2.4.8}$$

Therefore, by the use of (2.4.3), we get

$$\begin{aligned} d(R_1) + d(R_2) - 2d(N_s) &= d_s(R_1) + d_s(R_2) \\ &= \varepsilon_0(n_s(R_1) + n_s(R_2)) \\ &= \varepsilon_0(\#R_1 + \#R_2 - 2\#N_s) \\ &\leq \varepsilon_0(d(R_1) + (N-n) + d(R_2) + (N-n) - 2\#N_s) \\ &= \varepsilon_0((d(R_1) + d(R_2) - 2d(N_s)) + 2N - 2n - 2\#N_s + 2d(N_s)). \end{aligned}$$

Since $d(N_s) \leq \#N_s$, we have

$$\begin{aligned} d(R_1) + d(R_2) - 2d(N_s) &\leq \frac{\varepsilon_0}{1-\varepsilon_0}(2N - 2n - 2\#N_s + 2d(N_s)) \\ &< \frac{n+1-d(N_s)}{2N-2n+d(N_s)-\#N_s}(2N - 2n - 2\#N_s + 2d(N_s)) \\ &\leq n + 1 - d(N_s) \end{aligned}$$

by the use of (2.4.6). This implies that

$$d(R_1) + d(R_2) - d(N_s) \leq n.$$

By virtue of (2.4.2) we obtain

$$d(R_1 \cup R_2) \leq d(R_1) + d(R_2) - d(R_1 \cap R_2) \leq d(R_1) + d(R_2) - d(N_s) \leq n.$$

Therefore, $R_1 \cup R_2 \in \mathcal{R}$. By the definition of ε_0 we see

$$\varepsilon_0 \leq \varepsilon_s(R_1 \cup R_2). \tag{2.4.9}$$

In this situation, it holds that

$$d_s(R_1 \cap R_2) \geq \varepsilon_0 n_s(R_1 \cap R_2). \tag{2.4.10}$$

In fact, if $d(R_1 \cap R_2) > d(N_s)$, then $R_1 \cap R_2 \in \mathcal{R}$ and so (2.4.10) is an immediate consequence of the definition of ε_0. If $\#(R_1 \cap R_2) = \#N_s$, then

$R_1 \cap R_2 = N_s$. In this case, (2.4.10) is also true because $n_s(R_1 \cap R_2) = 0$. Assume that $d(R_1 \cap R_2) = d(N_s)$ and $\#(R_1 \cap R_2) > \#N_s$. Then we have necessarily $s > 0$, and

$$d(R_1 \cap R_2) - d(N_{s-1}) = d_{s-1}(N_s) > 0$$

because $\varepsilon_{N_{s-1}}(N_s) > 0$. This leads to an absurd conclusion:

$$\begin{aligned}\varepsilon_{s-1}(N_s) &\leq \varepsilon_{s-1}(R_1 \cap R_2)\\ &= \frac{d(R_1 \cap R_2) - d(N_{s-1})}{\#(R_1 \cup R_2) - \#N_{s-1}} < \frac{d(N_s) - d(N_{s-1})}{\#N_s - \#N_{s-1}} = \varepsilon_{s-1}(N_s).\end{aligned}$$

Thus, we have shown (2.4.10) for all possible cases.

By (2.4.2), (2.4.8) and (2.4.10), we obtain

$$\begin{aligned}\varepsilon_s(R_1 \cup R_2) &= \frac{d_s(R_1 \cup R_2)}{n_s(R_1 \cup R_2)}\\ &\leq \frac{d_s(R_1) + d_s(R_2) - d_s(R_1 \cap R_2)}{n_s(R_1) + n_s(R_2) - n_s(R_1 \cap R_2)}\\ &\leq \varepsilon_0.\end{aligned}$$

Combining this with (2.4.9), we can conclude $\varepsilon_s(R_1 \cup R_2) = \varepsilon_0$, which gives (2.4.7).

To complete the proof of Proposition 2.4.4, we define

$$N_{s+1} = \bigcup \{R; R \in \tilde{\mathcal{R}}\},$$

which is contained in $\tilde{\mathcal{R}}$ by (2.4.7). Then, for the family $\{N_i; 0 \leq i \leq s+1\}$, condition (i) is obviously satisfied, and conditions (ii) and (iii) are also satisfied by (2.4.5), (2.4.6) and the definition of N_{s+1}. The proof of Proposition 2.4.4 is completed.

THEOREM 2.4.11. *Let $H_1, H_2, \ldots, H_q$ be hyperplanes in $P^n(\mathbf{C})$ located in N-subgeneral position, where $q > 2N - n + 1$. Then there are some constants $\omega(1), \ldots, \omega(q)$ and θ satisfying the following conditions:*

(i) $0 < \omega(j) \leq \theta \leq 1 \qquad (1 \leq j \leq q)$,

(ii) $\sum_{j=1}^{q} \omega(j) = \theta(q - 2N + n - 1) + n + 1$,

(iii) $\dfrac{n+1}{2N-n+1} \leq \theta \leq \dfrac{n+1}{N+1}$,

(iv) *if* $R \subset Q, 0 < \#R \leq N+1$, *then* $\sum_{j \in R} \omega(j) \leq d(R)$.

DEFINITION 2.4.12. We call constants $\omega(j)(1 \leq j \leq q)$ and θ with the properties (i) $\sim$ (iv) *Nochka weights* and a *Nochka constant* for hyperplanes $H_1, \ldots, H_q$ respectively.

PROOF OF THEOREM 2.4.11. If $N = n$, then $\omega(j) := 1$ $(1 \leq j \leq q)$ and $\theta := 1$ satisfy the conditions (i) $\sim$ (iv). Assume that $N > n$. Take a family $\{N_i; 0 \leq i \leq s\}$ satisfying conditions (i) $\sim$ (iv) of Proposition 2.4.4. We then have $\#N_s \leq N$ by Proposition 2.4.4, (i) and the N-nondegeneracy of H_j's. Take some subset N_{s+1} such that $N_s \subseteq N_{s+1} \subseteq Q$ and $\#N_{s+1} = 2N - n + 1$. We then have $d(N_{s+1}) = n + 1$. As in the proof of Proposition 2.4.4, we use the abbreviated notation $d_i(R) = d_{N_i}(R), \varepsilon_i(R) = \varepsilon_{N_i}(R)$ and $n_i(R) = \#R - \#N_i$ for every R with $N_i \subseteq R \subseteq Q$. We set

$$\theta := \varepsilon_s(N_{s+1}) = \frac{n+1-d(N_s)}{2N-n+1-\#N_s}$$

and

$$\omega(j) := \begin{cases} \varepsilon_i(N_{i+1}) & \text{if } j \in N_{i+1} - N_i \text{ for some } i \text{ with } 0 \leq i \leq s, \\ \theta & \text{if } j \notin N_{s+1}. \end{cases}$$

We shall prove that these $\omega(j)$ and θ satisfy all conditions of Theorem 2.4.11.

We can easily check $\theta = \varepsilon_s(N_{s+1}) \leq 1$ by the use of (2.4.3). Take an arbitrary index $j \in Q$. If $j \notin N_s$, then (i) is obvious. For $j \in N_s$, there is one and only one i with $j \in N_{i+1} - N_i$. Then, by condition (ii) of Proposition 2.4.4 we have $\omega(j) = \varepsilon_i(N_{i+1}) < \theta$ and so (i) holds.

We next check condition (ii). The set Q is written as the disjoint union of $Q - N_{s+1}$ and $N_i - N_{i-1} (1 \leq i \leq s+1)$. Therefore,

$$\begin{aligned} \sum_{j=1}^{q} \omega(j) &= \sum_{j \in Q - N_{s+1}} \omega(j) + \sum_{i=1}^{s+1} \sum_{j \in N_i - N_{i-1}} \omega(j) \\ &= \theta(q - 2N + n - 1) + \sum_{i=1}^{s+1} (d(N_i) - d(N_{i-1})) \\ &= \theta(q - 2N + n - 1) + d(N_{s+1}) \\ &= \theta(q - 2N + n - 1) + n + 1. \end{aligned}$$

The assertion (iii) is now obvious. In fact, we have by (i) and (ii)

$$n+1=\sum_{j=1}^{q}\omega(j)-\theta(q-2N+n-1)\leq\theta(2N-n+1),$$

and by (2.4.3)

$$\theta=\frac{n+1-d(N_s)}{N+1+(N-n-\#N_s)}\leq\frac{n+1-d(N_s)}{N+1-d(N_s)}\leq\frac{n+1}{N+1}.$$

Finally we prove the assertion (iv). Take an arbitrary set R with $0<\#R\leq N+1$. We first consider the case where $d(R\cup N_s)=n+1$. Then, since $\#R\leq d(R)+N-n$ by (2.4.3) and

$$n+1-d(N_s)=d(N_s\cup R)-d(N_s)\leq d(R)$$

by (2.4.2), the assertion (i) implies that

$$\begin{aligned}\sum_{j\in R}\omega(j)&\leq\theta\#R\leq\theta(d(R)+N-n)\\&\leq\theta d(R)\left(1+\frac{N-n}{d(R)}\right)\\&\leq\theta d(R)\left(1+\frac{N-n}{n+1-d(N_s)}\right)\\&=d(R)\frac{N+1-d(N_s)}{2N-n+1-\#N_s}\\&\leq d(R).\end{aligned}$$

Therefore, the assertion (iv) holds in this case.

Assume that $d(R\cup N_s)\leq n$. Then $\#(N_s\cup R)\leq N$ by the assumption of N-nondegeneracy of the H_j's. We set

$$R_i:=\begin{cases}R\cap N_i & \text{if}\quad 0\leq i\leq s,\\ R & \text{if}\quad i=s+1.\end{cases}$$

We show next the following:

(2.4.13) *For* $1\leq i\leq s+1$, *if* $\#R_i>\#R_{i-1}$, *then* $d(R_i\cup N_{i-1})>d(N_{i-1})$.

For $i=1$, $d(R_1\cup N_0)=d(R_1)>0=d(N_0)$ by the assumption. Let $i>1$. Assume that $d(R_i\cup N_{i-1})=d(N_{i-1})$. Then $d_{i-2}(R_i\cup$

$N_{i-1}) = d(N_{i-1}) - d(N_{i-2}) > 0$. By Proposition 2.4.4, (iii), $\varepsilon_{i-2}(N_{i-1}) \leq \varepsilon_{i-2}(N_{i-1} \cup R_i)$ holds. On the other hand,

$$\begin{aligned}\varepsilon_{i-2}(N_{i-1} \cup R_i) &= \frac{d(N_{i-1} \cup R_i) - d(N_{i-2})}{\#(N_{i-1} \cup R_i) - \#(N_{i-2})} \\ &\leq \frac{d(N_{i-1}) - d(N_{i-2})}{\#N_{i-1} - \#N_{i-2}} = \varepsilon_{i-2}(N_{i-1}).\end{aligned}$$

Therefore, by Proposition 2.4.4, (iii), we have $\#R_i = \#R_{i-1}$, which contradicts the assumption.

Moreover, we can prove that

$$(2.4.14) \quad (\#R_i - \#R_{i-1})\varepsilon_{i-1}(N_i) \leq d(R_i) - d(R_{i-1}) \ (1 \leq i \leq s+1).$$

To show this, we may assume that $\#R_i > \#R_{i-1}$. Then by (2.4.13) we have $d(N_{i-1} \cup R_i) > d(N_{i-1})$. Hence, we have

$$\varepsilon_{i-1}(N_i) \leq \varepsilon_{i-1}(R_i \cup N_{i-1})$$

by Proposition 2.4.4, (iii) for the case $1 \leq i \leq s$ and (iv) for the case $i = s+1$. Since $\#(R_i \cup N_{i-1}) = \#N_{i-1} + \#R_i - \#(R_i \cap N_{i-1})$ and

$$d(R_i \cup N_{i-1}) \leq d(N_{i-1}) + d(R_i) - d(R_i \cap N_{i-1})$$

by (2.4.3), we have

$$\begin{aligned}\varepsilon_{i-1}(N_i) &\leq \varepsilon_{i-1}(R_i \cup N_{i-1}) \\ &= \frac{d(R_i \cup N_{i-1}) - d(N_{i-1})}{\#(R_i \cup N_{i-1}) - \#N_{i-1}} \leq \frac{d(R_i) - d(R_{i-1})}{\#R_i - \#R_{i-1}}.\end{aligned}$$

This gives (2.4.14).

The proof of (iv) is now easy. In fact, as a result of (2.4.13) and (2.4.14), we can conclude

$$\begin{aligned}\sum_{j \in R} \omega(j) &= \sum_{i=1}^{s+1} \sum_{j \in R_i - R_{i-1}} \omega(j) \\ &\leq d(R) - d(R_s) + \sum_{i=1}^{s} d(R_i) - d(R_{i-1}) \\ &= d(R).\end{aligned}$$

Thus, the proof of Theorem 2.4.11 is completed.

Related to Theorem 2.4.11, we give the following proposition for later use.

PROPOSITION 2.4.15. *Let $H_1, \dots, H_q$ be hyperplanes in $P^n(\mathbf{C})$ located in N-subgeneral position and let $\omega(1), \dots, \omega(q)$ be Nochka weights for them, where $q > 2N - n + 1$. Consider an arbitrary subset R of $Q := \{1, 2, \dots, q\}$ with $0 < \#R \leq N + 1$ and $d + 1 = d(R)$, and arbitrary real constants $E_1, \dots, E_q$ not less than one. Then, there are some $j_0, \dots, j_d \in Q$ such that $H_{j_0}, \dots, H_{j_d}$ are linearly independent and*

$$\prod_{j \in R} E_j^{\omega(j)} \leq \prod_{\ell=0}^{d} E_{j_\ell}.$$

PROOF. Changing indices, we may assume that

$$E_1 \geq E_2 \geq \dots \geq E_q.$$

Taking nonzero vectors $A_j \in \mathbf{C}^{n+1}$, we represent the given hyperplanes H_j as

$$H_j : \langle W, A_j \rangle = 0 \qquad (1 \leq j \leq q).$$

For a set $R \subseteq Q$ we set $V(R) := \{\{A_j; j \in R\}\}$ and use the previous notation $d(R) := \dim V(R)$.

We shall choose j_ℓ's by induction on ℓ. We first set $j_0 := 1, R_0 := \{j_0\}$ and $S_0 := \{j \in R; A_j \in V(R_0)\}$. Next, choose

$$j_1 := \min\{j \in R; A_j \notin V(R_0)\}$$

and set $R_1 := \{j_0, j_1\}, S_1 := \{j \in R - S_0; A_j \in V(R_1)\}$. Similarly, choose

$$j_2 := \min\{j \in R; A_j \notin V(R_1)\}$$

and set $R_2 := \{j_0, j_1, j_2\}, S_2 := \{j \in R - S_1; A_j \in V(R_2)\}$. We can repeat this processes until $V(R_d) = V(R)$ for $d = \dim V(R) - 1$, because $A_{j_0}, \dots, A_{j_\ell}$ are linearly independent for each ℓ. Then R is represented as a disjoint union

$$R = S_0 \cup S_1 \cup \dots \cup S_d$$

and $E_j \leq E_{j_\ell}$ holds for each $j \in S_\ell$. Therefore, we obtain

$$\begin{aligned} \prod_{j \in R} E_j^{\omega(j)} &= \prod_{\ell=0}^{d} \prod_{j \in S_\ell} E_j^{\omega(j)} \\ &\leq \prod_{\ell=0}^{d} \prod_{j \in S_\ell} E_{j_\ell}^{\omega(j)} = \prod_{\ell=0}^{d} E_{j_\ell}^{\Sigma_{j \in S_\ell} \omega(j)}. \end{aligned}$$

For convenience's sake, we set $T_\ell := S_0 \cup S_1 \cup \ldots \cup S_\ell$. holds. Then, $\sum_{j \in T_\ell} \omega(j) \leq d(T_\ell) = \ell$ holds. It follows that

$$\begin{aligned}
\prod_{j \in R} E_j^{\omega(j)} &= E_{j_0} E_{j_0}^{-1+\Sigma_{j \in T_0}\omega(j)} \prod_{\ell=1}^{d} E_{j_\ell}^{\Sigma_{j \in S_\ell}\omega(j)} \\
&\leq E_{j_0} E_{j_1}^{-1+\Sigma_{j \in T_0}\omega(j)} \prod_{\ell=1}^{d} E_{j_\ell}^{\Sigma_{j \in S_\ell}\omega(j)} \\
&\leq E_{j_0} E_{j_1} E_{j_1}^{-2+\Sigma_{j \in T_1}\omega(j)} \prod_{\ell=2}^{d} E_{j_\ell}^{\Sigma_{j \in S_\ell}\omega(j)} \\
&\leq E_{j_0} E_{j_1} E_{j_2} E_{j_2}^{-3+\Sigma_{j \in T_2}\omega(j)} \prod_{\ell=2}^{d} E_{j_\ell}^{\Sigma_{j \in S_\ell}\omega(j)}.
\end{aligned}$$

Repeating this processes, we obtain

$$\prod_{j \in R} E_j^{\omega(j)} \leq E_{j_0} E_{j_1} \ldots E_{j_d} E_{j_d}^{-d+\Sigma_{j \in T_d}\omega(j)}.$$

On the other hand, since $E_{j_d} \geq 1$ and

$$\sum_{j \in T_d} \omega(j) \leq d(T_d) = d(R) = d,$$

we have $E_{j_d}^{-d+\Sigma_{j \in T_0}\omega(j)} \leq 1$. This gives the desired inequality.

§2.5 Sum to product estimates for holomorphic curves

As in §2.4, we consider contact functions for a nondegenerate holomorphic map of an open Riemann surface M into $P^n(\mathbf{C})$.

The following inequality is a generalization of Lemma 1.4.8 to the case of holomorphic curves in $P^n(\mathbf{C})$.

PROPOSITION 2.5.1. *For each positive ε there exists a constant $a_0(\varepsilon)$, depending only on ε, such that for any hyperplane H in $P^n(\mathbf{C})$ and any constant $a \geq a_0(\varepsilon)$*

$$dd^c \log \frac{1}{\log(a/\varphi_k(H))} \geq \frac{\varphi_{k+1}(H)}{\varphi_k(H) \log^2(a/\varphi_k(H))} \Omega_k - \varepsilon \Omega_k.$$

PROOF. By the use of Theorems 2.3.9 and 2.3.11, we obtain

$$\begin{aligned}
dd^c \log\frac{1}{\log(a/\varphi_k)} &= \frac{dd^c\log\varphi_k}{\log(a/\varphi_k)} + \frac{d\varphi_k\wedge d^c\varphi_k}{\varphi_k^2\log^2(a/\varphi_k)} \\
&= \frac{(\varphi_{k-1}\varphi_{k+1}-\varphi_k^2)}{\varphi_k^2\log(a/\varphi_k)}\Omega_k + \frac{(\varphi_{k+1}-\varphi_k)(\varphi_k-\varphi_{k-1})}{\varphi_k^2\log^2(a/\varphi_k)}\Omega_k \\
&= \frac{\varphi_{k+1}}{\varphi_k\log^2(a/\varphi_k)}\Omega_k + \frac{\varphi_{k-1}\varphi_{k+1}}{\varphi_k^2}\left(\frac{1}{\log(a/\varphi_k)} - \frac{1}{\log^2(a/\varphi_k)}\right)\Omega_k \\
&\quad + \frac{\varphi_{k-1}}{\varphi_k\log^2(a/\varphi_k)}\Omega_k - \left(\frac{1}{\log(a/\varphi_k)} + \frac{1}{\log^2(a/\varphi_k)}\right)\Omega_k.
\end{aligned}$$

Since $\varphi_k(H)\leq 1$ and since the second and the third terms of the right hand side of the above identities are nonnegative for $a>e$, if we choose a constant $a_0(\epsilon)$ such that $a_0(\varepsilon)>e$ and

$$\frac{1}{\log a_0(\varepsilon)} + \frac{1}{\log^2 a_0(\varepsilon)} < \varepsilon,$$

it satisfies the desired inequality.

We now give the so-called "sum to product estimate" for the case of holomorphic curves in $P^n(\mathbf{C})$.

THEOREM 2.5.2. *Let $H_1,\dots,H_q$ $(q>2N-n+1)$ be hyperplanes in $P^n(\mathbf{C})$ located in N-subgeneral position and let $\omega(j)(1\leq j\leq q)$ be Nochka weights for these hyperplanes. For an arbitrarily given $a>1$ and $0\leq k\leq n-1$ we set*

$$\Phi_{jk} = \frac{\varphi_{k+1}(H_j)}{\varphi_k(H_j)\log^2(a/\varphi_k(H_j))}.$$

Then, there exists a positive constant C_k depending only on k and $H_j(1\leq j\leq q)$ such that

$$\sum_{j=1}^{q}\omega(j)\Phi_{jk} \geq C_k\left(\prod_{j=1}^{q}\Phi_{jk}^{\omega(j)}\right)^{1/(n-k)}$$

holds on $M-\bigcup_{1\leq j\leq q}\{z;\varphi_k(H_j)(z)=0\}$.

PROOF. Let

$$H_j:\ \langle W, A_j\rangle = 0 \qquad (1\leq j\leq q),$$

where we choose A_j with $|A_j| = 1$ for each j. By the assumption, $A_1, \dots, A_q$ are located in N-subgeneral position.

Now, we consider the set $\mathcal{R}_k$ of all subsets R of $Q := \{1, 2, \dots, q\}$ such that $d(R) \leq n-k$. For each element P in $G(n,k)$ we take a decomposable $(k+1)$-vector E such that

$$P = \{X \in \mathbf{C}; E \wedge X = 0\}$$

and set

$$\psi_k(P) := \max_{R \in \mathcal{R}_k} \min \left\{ \frac{|E \vee A_j|^2}{|E|^2}; j \notin R \right\}.$$

Obviously, $\psi_k(P)$ depends only on P. Therefore, we can regard ψ_k as a function on the Grassmann manifold $G(n,k)$. For each nonzero $(k+1)$-vector $E = E_0 \wedge \dots \wedge E_k$ we set

$$R = \{j \in Q; E \vee A_j = 0\}.$$

Then, since $E \vee A_j = 0$ means that A_j is contained in the orthogonal complement of the vector space $\{\{E_0, \dots, E_k\}\}$ according to (2.3.5), we see

$$d(R) = \dim\{\{A_j; j \in R\}\} \leq n - k,$$

namely, $R \in \mathcal{R}_k$. This yields that ψ_k is positive everywhere on $G(n,k)$. Since ψ_k is obviously continuous and $G(n,k)$ is compact, we can take a positive constant δ such that $\psi_k(P) > \delta$ for each $P \in G(n,k)$.

Take an arbitrary point z with $F_k(z) \not\equiv 0$. The vector space generated by $F(z), F'(z), \dots, F^{(k)}(z)$ determines a point in $G(n,k)$. Therefore, there is a set R in $\mathcal{R}_k$ with $d(R) \leq n-k$ such that $\varphi_k(H_j)(z) \geq \delta$ for all $j \notin R$. Then, we can choose a finite positive constant K depending only on H_j such that $\Phi_{jk}(z) \leq K$ for all $j \notin R$. Set

$$S := \{j; \Phi_{jk} > K\}, \ \ell := \sum_{j \in S} \omega(j).$$

We may assume $S \neq \emptyset$. In fact, if S is an empty set, by the use of (1.4.6)

we have

$$\sum_{j=1}^{q}\omega(j)\Phi_{jk}\geq\left(\sum_{j=1}^{q}\omega(j)\right)\left(\prod_{j=1}^{q}\Phi_{jk}^{\omega(j)}\right)^{\frac{1}{\Sigma_j\omega(j)}}$$

$$\geq(n+1)K\left(\prod_{j=1}^{q}\left(\frac{\Phi_{jk}}{K}\right)^{\frac{\omega(j)}{n-k}}\right)^{\frac{n-k}{\Sigma_j\omega(j)}}$$

$$\geq(n+1)K\left(\prod_{j=1}^{q}\left(\frac{\Phi_{jk}}{K}\right)^{\omega(j)}\right)^{\frac{1}{n-k}},$$

because $n+1\leq\sum_{j=1}^{q}\omega(j)$. We see $S\subseteq R$ and so $d(S)\leq n-k$ holds. In this situation, $\ell\leq n-k$ by Theorem 2.4.11, (iv) and $\#S<N+1$ because of N-nondegeneracy of H_j. By the help of (1.4.6) we obtain

$$\sum_{j=1}^{q}\omega(j)\Phi_{jk}\geq\sum_{j\in S}\omega(j)\Phi_{jk}\geq K\ell\prod_{j\in S}\left(\frac{\Phi_{jk}}{K}\right)^{\omega(j)/\ell}$$

$$\geq K\ell\prod_{j\in S}\left(\frac{\Phi_{jk}}{K}\right)^{\omega(j)/(n-k)}\geq C\prod_{j=1}^{q}\left(\frac{\Phi_{jk}}{K}\right)^{\omega(j)/(n-k)}$$

for some constant $C>0$. Thus, we can find a positive constant C_k depending only on k and $H_1,\dots,H_q$ which satisfies the desired inequality in Theorem 2.5.2.

Now, we give the following theorem, which plays a fundamental role in the following sections.

THEOREM 2.5.3. *Let $H_1,\dots,H_q$ be hyperplanes in $P^n(\mathbf{C})$ located in N-subgeneral position and let $\omega(j)\,(1\leq j\leq q)$ and θ be Nochka weights and a Nochka constant for these hyperplanes. For every $\varepsilon>0$ there exist some positive numbers $a(>1)$ and C, depending only on ε and $H_j(1\leq j\leq q)$, such that*

(2.5.4)

$$dd^c\log\frac{\prod_{k=0}^{n-1}|F_k|^{2\varepsilon}}{\prod_{1\leq j\leq q,0\leq k\leq n-1}\log^{2\omega(j)}(a/\varphi_k(H_j))}$$

$$\geq C\left(\frac{|F_0|^{2\theta(q-2N+n-1)}|F_n|^2}{\prod_{j=1}^{q}(|F(H_j)|^2\prod_{k=0}^{n-1}\log^2(a/\varphi_k(H_j)))^{\omega(j)}}\right)^{\frac{2}{n(n+1)}}dd^c|z|^2.$$

PROOF. We denote the left hand side of (2.5.4) by A. Then, by the definition of Ω_k, it may be rewritten as

$$A = \varepsilon \sum_{k=0}^{n-1} \Omega_k + \sum_{j=1}^{q} \omega(j) \sum_{k=0}^{n-1} dd^c \log \frac{1}{\log^2(a/\varphi_k(H_j))}.$$

Choose a positive number $a_0(\varepsilon/\ell)$ with properties as in Proposition 2.5.1, where $\ell = \sum_{j=1}^{q} \omega(j)$. For an arbitrarily fixed $a \geq a_0(\varepsilon/\ell)$ we obtain

$$\begin{aligned} A &\geq \varepsilon \sum_{k=0}^{n-1} \Omega_k + \sum_{j=1}^{q} \omega(j) \sum_{k=0}^{n-1} \left(\frac{2\varphi_{k+1}(H_j)}{\varphi_k(H_j) \log^2(a/\varphi_k(H_j))} - \frac{\varepsilon}{\ell} \right) \Omega_k \\ &= \sum_{k=0}^{n-1} 2 \left(\sum_{j=1}^{q} \omega(j) \Phi_{jk} \right) \Omega_k, \end{aligned}$$

where Φ_{jk} is the quantity defined in Theorem 2.5.2. Then, by virtue of Theorem 2.5.2,

$$A \geq C_1 \sum_{k=0}^{n-1} 2 \left(\prod_{j=1}^{q} \Phi_{jk}^{\omega(j)} \right)^{\frac{1}{n-k}} \Omega_k$$

for a constant $C_1 > 0$. We now apply the inequality (1.4.6) to $a_k := n-k$ and $x_k := \prod_{j=1}^{q} \Phi_{jk}^{\omega(j)} h_k^{n-k}$ to see

$$A \geq C_2 \prod_{k=0}^{n-1} \left(h_k^{n-k} \prod_{j=1}^{q} \Phi_{jk}^{\omega(j)} \right)^{\frac{2}{n(n+1)}} dd^c|z|^2$$

for some $C_2 > 0$, where $\Omega_k = h_k dd^c|z|^2$. On the other hand, we have

$$\begin{aligned} \prod_{k=0}^{n-1} \Phi_{jk} &= \prod_{k=0}^{n-1} \frac{\varphi_{k+1}(H_j)}{\varphi_k(H_j)} \frac{1}{\log^2(a/\varphi_k(H_j))} \\ &= \frac{|F_0|^2}{|F(H_j)|^2} \prod_{k=0}^{n-1} \frac{1}{\log^2(a/\varphi_k(H_j))} \end{aligned}$$

and

$$\prod_{k=0}^{n-1} h_k^{n-k} = \prod_{k=0}^{n-1} \left(\frac{|F_{k-1}|^2 |F_{k+1}|^2}{|F_k|^4} \right)^{n-k} = \frac{|F_n|^2}{|F_0|^{2(n+1)}},$$

because $\varphi_0(H_j) = |F(H_j)|^2/|F|^2, \varphi_n(H_j) = 1$ and the products telescope. Therefore, we get

$$A \geq C \left(\frac{|F|^{2(\ell-n-1)}|F_n|^2}{\prod_{j=1}^q (|F(H_j)|^2 \prod_{k=0}^{n-1} \log^2(a/\varphi_k(F_j)))^{\omega(j)}} \right)^{\frac{2}{n(n+1)}} dd^c|z|^2.$$

Since $\ell - n - 1 = \theta(q - 2N + n - 1)$ by Theorem 2.4.11, (ii), this concludes the proof of Theorem 2.5.3.

COROLLARY 2.5.5. *Take holomorphic functions g_k with $\nu_{g_k} = \nu_{|F_k|}$ and set $\tilde{F}_k = F_k/g_k$ for each $k = 0, 1, \ldots, n-1$. Set*

$$\hat{h} = \frac{\prod_{k=0}^{n-1} |\tilde{F}_k|^{2\varepsilon}}{\prod_{k=0}^{n-1} \prod_{j=1}^q \log^{2\omega(j)}(a/\varphi_k(H_j))}$$

for positive numbers ε and a. Then, if we choose a suitable a, there exists a positive constants C such that

$$dd^c \log \hat{h} \geq$$

$$C \left(\frac{|F_0|^{2\theta(q-2N+n-1)}|F_n|^2 \hat{h}}{|\tilde{F}_0|^{2\varepsilon} \cdots |\tilde{F}_{n-1}|^{2\varepsilon} \prod_{j=1}^q |F(H_j)|^{2\omega(j)}} \right)^{\frac{2}{n(n+1)}} dd^c|z|^2. \tag{2.5.6}$$

PROOF. Since $dd^c \log |F(H_j)|^2 = dd^c \log |g_k|^2 = 0$, the term $dd^c \log \hat{h}$ is bounded from below by the right hand side of (2.5.4). The inequality (2.5.6) is an immediate consequence of (2.5.4).

For later use, we shall prove here another proposition.

PROPOSITION 2.5.7. *Set $\sigma_n := n(n+1)/2$ and $\tau_n = \sum_{k=1}^n \sigma_k$. Then,*

$$dd^c \log |F_0|^2 |F_1|^2 \cdots |F_{n-1}|^2 \geq \frac{\tau_n}{\sigma_n} \left(\frac{|F_0|^2 \cdots |F_n|^2}{|F_0|^{2\sigma_{n+1}}} \right)^{1/\tau_n} dd^c|z|^2.$$

PROOF. Since $dd^c \log |F_k|^2$ $(0 \leq k \leq n-1)$ are nonnegative, we obtain by Proposition 2.2.11

$$\sigma_n dd^c \log |F_0|^2 \cdots |F_{n-1}|^2 \geq \sum_{k=0}^{n-1} \sigma_k \frac{|F_{k-1}|^2 |F_{k+1}|^2}{|F_k|^4} dd^c|z|^2.$$

Apply (1.4.6) to the numbers $a_k = \sigma_{n-k}$ and $x_k = |F_{k-1}|^2|F_{k+1}|^2/|F_k|^4$ to see

$$\begin{aligned}\sigma_n dd^c \log |F_0|^2 \cdots |F_{n-1}|^2 \\ &\geq \tau_n \left(\prod_{k=0}^{n-1} \left(\frac{|F_{k-1}|^2|F_{k+1}|^2}{|F_k|^4} \right)^{\sigma_{n-k}} \right)^{1/\tau_n} dd^c|z|^2 \\ &= \tau_n \left(\frac{|F_1|^2 \cdots |F_{n-1}|^2|F_n|^2}{|F_0|^{n^2+3n}} \right)^{1/\tau_n} dd^c|z|^2,\end{aligned}$$

where $|F_{-1}| \equiv 1$. This gives Proposition 2.5.7.

§2.6 Contracted curves

In this section, we state some properties of contracted curves of derived curves, which are used in the proof of the defect relation for derived curves. Readers who are interested only in minimal surfaces may skip this section.

Let $f : \mathbf{C} \to P^n(\mathbf{C})$ be a non-degenerate holomorphic curve with a reduced representation $f = (f_0 : \cdots : f_n)$. As usual, we set $F := (f_0, \ldots, f_n)$ and define the map

$$F_k := F \wedge F' \wedge \ldots \wedge F^{(k)} : C \to \bigwedge^{k+1} \mathbf{C}^{n+1}$$

for $k := 1, \ldots, n$. Take a nonzero decomposable $(k-1)$-vector $A^{k-1} = A_0 \wedge \ldots \wedge A_{k-1} \in \bigwedge^k \mathbf{C}^{n+1}$, and set

$$F_{A^{k-1}} := F_k \vee A^{k-1} \in \mathbf{C}^{n+1}.$$

We then have $F_{A^{k-1}} \not\equiv 0$. In fact, if we take some vector A_k with $A^{k-1} \wedge A_k \neq 0$, then

$$\begin{aligned}\langle F_{A^{k-1}}, A_k \rangle &= \langle F_k, A_0 \wedge \ldots \wedge A_k \rangle \\ &= \det(\langle F^{(i)}, A_j \rangle; 0 \leq i, j \leq k) \not\equiv 0,\end{aligned}$$

because $\langle F, A_0 \rangle, \ldots, \langle F, A_k \rangle$ are linearly independent. Consider the vector space $V(A^{k-1}) = \{X \in \mathbf{C}^{n+1}; A^{k-1} \wedge X = 0\}$ and define

$$A^\perp := V(A^{k-1})^\perp = \{X; \langle A_j, X \rangle = 0 \text{ for } j = 0, 1, \ldots, k-1\}.$$

Since $\langle F_{A^{k-1}}, X\rangle = \langle F_k, A^{k-1} \wedge X\rangle = 0$ for every $X \in V(A^{k-1})$, the image of $F_{A^{k-1}}$ is contained in $A^{\perp}$. By $P(A^{\perp})$ we denote the projective subspace of $P^n(\mathbf{C})$ associated with the vector subspace $A^{\perp}$ of $\mathbf{C}^{n+1}$.

DEFINITION 2.6.1. We call the holomorphic curve in $P(A^{\perp})$ defined by the map $F_{A^{k-1}}$ the *contracted curve* for f to the direction A^{k-1} and denote it by $f_{A^{k-1}}$.

Here, we note that $f_{A^{k-1}} : \mathbf{C} \to P(A^{\perp})$ is nondegenerate because of the following:

PROPOSITION 2.6.2. *In the above situation, it holds that*

$$(F_{A^{k-1}})_\ell := F_{A^{k-1}} \wedge F'_{A^{k-1}} \wedge \ldots \wedge F^{(\ell)}_{A^{k-1}} = \langle F_{k-1}, A^{k-1}\rangle^{\ell} F_{k+\ell} \vee A^{k-1}$$

for $0 \leq \ell \leq n-k$.

PROOF. The proof is given by induction on k. For the case $k = 1$, we prove by induction on ℓ the following:

(2.6.3) $(F \wedge F' \vee A)_\ell = \langle F, A\rangle^{\ell} F_{\ell+1} \vee A$ *for every* $A \in \mathbf{C}^{n+1}$ *and* $0 \leq \ell \leq n-1$.

It suffices to prove (2.6.3) at every point z_0 with $\langle F(z_0), A\rangle \neq 0$ by the continuity of both sides. Here, we may replace the vector-valued holomorphic function F by $\tilde{F} := hF$ for a nonzero holomorphic function h. Because, by this replacement, both sides of (2.6.3) are multiplied by the same factor $h^{2\ell+2}$. We choose a nonzero holomorphic function h in a neighborhood of z_0 such that, for $\tilde{F} := hF$,

$$\langle \tilde{F}^{(\ell)}(z_0), A\rangle = 0 \qquad \ell = 1, 2, \ldots, n.$$

This is possible. For, assume that there exists some polynomial

$$h(z) = 1 + \frac{c_1}{1!}(z - z_0) + \cdots + \frac{c_n}{n!}(z - z_0)^n$$

such that the function $\tilde{F} = hF$ satisfies the identities

$$\langle \tilde{F}^{(\ell)}(z_0), A\rangle \equiv \sum_{m=0}^{\ell} \binom{\ell}{m} c_m \langle F^{(\ell-m)}(z_0), A\rangle = 0$$

for $\ell = 1, 2, \ldots, n$, where $c_0 := 1$. Regard this as a system of simultaneous linear equations with unknown variables c_j's. It has a system of solutions because the determinant of the matrix of the coefficients take the nonzero value $\langle F(z_0), A\rangle^n$. Changing notation, we assume that F itself satisfies the condition

$$\langle F^{(\ell)}(z_0), A\rangle = 0 \tag{2.6.4}$$

for $\ell = 1, 2, \dots, n$. Then

$$F_\ell \vee A = \langle F, A\rangle F' \wedge \dots \wedge F^{(\ell)} \tag{2.6.5}$$

Because, by (2.6.4), for each decomposable $(\ell-1)$-vector $X = X_1 \wedge \dots \wedge X_\ell$ we have

$$\begin{aligned}
\langle F_\ell \vee A, X\rangle &= \langle F \wedge F' \wedge \dots \wedge F^{(\ell)}, A \wedge X\rangle \\
&= \det(\langle F^{(r)}, X_s\rangle; 0 \le r, s \le \ell) \\
&= \langle F, A\rangle \langle F' \wedge \dots \wedge F^{(\ell)}, X\rangle
\end{aligned}$$

at z_0 by (2.6.4), where $X_0 := A$.

We have nothing to prove for the case $\ell = 0$ of (2.6.3). Suppose that (2.6.3) holds for the cases $\le \ell - 1$. By (2.6.5), the induction assumption can be rewritten as

$$(F \wedge F' \vee A)_{\ell-1} = \langle F, A\rangle^{\ell-1} F_\ell \vee A = \langle F, A\rangle^\ell F' \wedge \dots \wedge F^{(\ell)}$$

at z_0. Using (2.6.4) again, we obtain

$$\begin{aligned}
(F \wedge F' \vee A)^{(\ell)} &= F \wedge F^{(\ell+1)} \vee A + \sum_{m=1}^{\ell} \binom{\ell}{m} F^{(m)} \wedge F^{(\ell+1-m)} \vee A \\
&= F \wedge F^{(\ell+1)} \vee A = \langle F, A\rangle F^{(\ell+1)}
\end{aligned}$$

and $F_{\ell+1} \vee A = \langle F, A\rangle F' \wedge \dots \wedge F^{(\ell+1)}$ at z_0. Therefore, we conclude

$$\begin{aligned}
(F \wedge F' \vee A)_\ell &= (F \wedge F' \vee A)_{\ell-1} \wedge (F \wedge F' \vee A)^{(\ell)} \\
&= \langle F, A\rangle^{\ell+1} F' \wedge \dots \wedge F^{(\ell+1)} \\
&= \langle F, A\rangle^\ell F_{\ell+1} \vee A.
\end{aligned}$$

The assertion (2.6.3) is proved.

Now, we assume that Proposition 2.6.2 holds for k. For arbitrarily given A^k we represent A^k as $A^k = A^{k-1} \wedge A_k$. By applying the induction assumption for the case $\ell = 1$, we have

$$F_{A^k} \equiv F_{k+1} \vee A^k = (F_{k+1} \vee A^{k-1}) \vee A_k = \langle F_{k-1}, A^{k-1}\rangle^{-1}(F_{A^{k-1}})_1 \vee A_k.$$

On the other hand, by applying the induction assumption to the curve $F_{A^{k-1}}$ for the case $k = 1$ we see

$$((F_{A^{k-1}})_1 \vee A_k)_\ell = ((F_{A^{k-1}})_{A_k})_\ell = \langle F_{A^{k-1}}, A_k\rangle^\ell (F_{A^{k-1}})_{\ell+1} \vee A_k.$$

Since $\langle F_{A^{k-1}}, A_k\rangle = \langle F_k, A^k\rangle$, with the help of the induction assumption and the above identities we conclude

$$\begin{aligned}(F_{A^k})_\ell &= ((F_{A^{k-1}})_1 \vee A_k)_\ell \langle F_{k-1}, A^{k-1}\rangle^{-(\ell+1)} \\ &= \langle F_{A^{k-1}}, A_k\rangle^\ell (F_{A^{k-1}})_{\ell+1} \vee A_k \langle F_{k-1}, A^{k-1}\rangle^{-(\ell+1)} \\ &= \langle F_{A^{k-1}}, A_k\rangle^\ell F_{k+\ell+1} \vee A^{k-1} \vee A_k \\ &= \langle F_k, A^k\rangle^\ell F_{k+\ell+1} \vee A^k.\end{aligned}$$

This completes the proof of Proposition 2.6.2.

We shall describe some properties of the derived curves of contracted curves. To this end, we represent with the notation A^h $(0 \leq h \leq n-1)$ a decomposable unit h-vector in $\mathbf{C}^{n+1}$. Following [13], for a given A^h we define

$$\varphi_k(A^h) := \frac{|F_k \vee A^h|^2}{|F_k|^2} (\not\equiv 0), \quad \Omega_k(A^h) = dd^c \log |F_k \vee A^h|^2 \tag{2.6.6}$$

for $h \leq k$. For convenience's sake, we set $\varphi_k(A^h) = 0$ for $h > k$ and $\varphi_k(A^{-1}) \equiv 1$.

For a given A^k choose and fix a system of orthonormal vectors A_0, $A_1, \dots, A_k$ such that $A^k = A_0 \wedge \dots \wedge A_k$. In the followings, the notation $A^{h'} \subseteq A^{h''}$ means that

$$A^{h'} = A_{i_0} \wedge \dots \wedge A_{i_{h'}}, \quad A^{h''} = A_{j_0} \wedge \dots \wedge A_{j_{h''}}$$

for some indices $i_0 < \cdots < i_{h'}$ and $j_0 < \cdots < j_{h''}$ with $\{i_0, \cdots, i_{h'}\} \subseteq \{j_0, \cdots, j_{h''}\}$. We define

$$\psi_{\ell,h}(A^k) = \sum_{A^h \subseteq A^k} \frac{\varphi_\ell(A^h)}{\varphi_{\ell+1}(A^h)} \tag{2.6.7}$$

on the set $\{z; \varphi_{\ell+1}(A^h)(z) \neq 0 \text{ for any } A^h \subseteq A^k\}$. For convenience's sake, set $\psi_{\ell,-1} \equiv 1$. We can show the following:

PROPOSITION 2.6.8. *For each positive ε there exists a constant $a_0(\epsilon)$ such that for every A^k and $a \geq a_0(\varepsilon)$*

$$dd^c \log \prod_{A^h \subseteq A^{h+1} \subseteq A^k} \frac{1}{\log(a\varphi_\ell(A^h)/\varphi_\ell(A^{h+1}))} + \varepsilon \sum_{A^h \subseteq A^k} \Omega_\ell(A^h)$$
$$\geq (k-h)\frac{\psi_{\ell-1,h}(A^k)}{\psi_{\ell,h+1}(A^k)} \left(\prod_{A^h \subseteq A^{h+1} \subseteq A^k} \frac{1}{\log^2(a\varphi_\ell(A^h)/\varphi_\ell(A^{h+1}))} \right) \Omega_\ell$$

for $\ell > h, k > h$.

PROOF. In view of (2.6.6), Propositions 2.6.2 and 2.2.11, we have

$$\begin{aligned} \Omega_\ell(A^h) &= dd^c \log |(F_{A^h})_{\ell-h-1}|^2 \\ &= \frac{|(F_{A^h})_{\ell-h-2}|^2 |(F_{A^h})_{\ell-h}|^2}{|(F_{A^h})_{\ell-h-1}|^4} dd^c|z|^2 \\ &= \frac{|F_{\ell-1} \vee A^h|^2 |F_{\ell+1} \vee A^h|^2}{|F_\ell \vee A^h|^4} dd^c|z|^2 \qquad (2.6.9) \\ &= \frac{\varphi_{\ell-1}(A^h)\varphi_{\ell+1}(A^h)}{\varphi_\ell(A^h)^2}\Omega_\ell, \end{aligned}$$

where we used the identity $dd^c \log |\langle F_h, A^h \rangle|^2 = 0$.

Choose some h-vector A^h and $A^{h+1} := A^h \wedge A$ with $A^h \subseteq A^{h+1} \subseteq A^k$ and consider the contracted curve f_{A^h}. The k-th contact function of f_{A^h} and the pull-back of the Fubini-Study metric form are given by

$$\tilde{\varphi}_k(A) = \frac{|(F_{A^h})_k \vee A|^2}{|(F_{A^h})_k|^2}, \quad \tilde{\Omega}_k = dd^c \log |(F_{A^h})_k|^2$$

respectively. Then, by Proposition 2.6.2,

$$\tilde{\varphi}_k(A) = \frac{|(F_{h+k+1} \vee A^h) \vee A|^2}{|F_{h+k+1} \vee A^h|^2} = \frac{\varphi_{h+k+1}(A^{h+1})}{\varphi_{h+k+1}(A^h)} (\leq 1) \qquad (2.6.10)$$

and

$$\tilde{\Omega}_k = dd^c \log |F_{h+k+1} \vee A^h|^2 = \Omega_{h+k+1}(A^h).$$

Now, we apply Proposition 2.5.1 to the contracted curve. For an arbitrarily given $\epsilon > 0$ we can take a positive constant $a(\varepsilon)(\geq e)$ such that, for

$\ell > h$ and $a \geq a_0(\varepsilon)$,

$$
\begin{aligned}
(2.6.11)\qquad & dd^c \log \frac{1}{\log(a\varphi_\ell(A^h)/\varphi_\ell(A^{h+1}))} + \varepsilon\Omega_\ell(A^h) \\
& \geq \frac{(\varphi_{\ell+1}(A^{h+1})/\varphi_{\ell+1}(A^h))\Omega_\ell(A^h)}{(\varphi_\ell(A^{h+1})/\varphi_\ell(A^h))\log^2(a\varphi_\ell(A^h)/\varphi_\ell(A^{h+1}))} \\
& = \frac{\varphi_{\ell+1}(A^{h+1})\varphi_{\ell-1}(A^h)}{\varphi_\ell(A^{h+1})\varphi_\ell(A^h)} \frac{\Omega_\ell}{\log^2(a\varphi_\ell(A^h)\varphi_\ell(A^{h+1}))},
\end{aligned}
$$

where we used the identity (2.6.9). Sum (2.6.11) over all possible $A^h \subseteq A^{h+1} (\subseteq A^k)$ to obtain

(2.6.12)

$$
\begin{aligned}
& dd^c \log \prod_{A^h \subseteq A^{h+1}} \frac{1}{\log(a\varphi_\ell(A^h)/\varphi_\ell(A^{h+1}))} + \varepsilon \sum_{A^h \subseteq A^{h+1}} \Omega_\ell(A^h) \\
& \geq \sum_{A^h \subseteq A^{h+1}} \frac{\varphi_{\ell+1}(A^{h+1})\varphi_{\ell-1}(A^h)}{\varphi_\ell(A^{h+1})\varphi_\ell(A^h)} \left(\prod_{A^h \subseteq A^{h+1}} \frac{1}{\log^2\left(\frac{a\varphi_\ell(A^h)}{\varphi_\ell(A^{h+1})}\right)} \right) \Omega_\ell,
\end{aligned}
$$

where we used the fact:

$$\log(a\varphi_\ell(A^h)/\varphi_\ell(A^{h+1})) = \log(a/\tilde{\varphi}_{\ell-h-1}(A)) \geq 1.$$

Since for a fixed A^h there are $k-h$ A^{h+1}'s with $A^h \subseteq A^{h+1}$, we have

$$
\begin{aligned}
& \left(\sum_{A^h \subseteq A^{h+1}} \frac{\varphi_{\ell+1}(A^{h+1})\varphi_{\ell-1}(A^h)}{\varphi_\ell(A^{h+1})\varphi_\ell(A^h)} \right) \sum_{A^{h+1} \subseteq A^k} \frac{\varphi_\ell(A^{h+1})}{\varphi_{\ell+1}(A^{h+1})} \\
& \qquad \geq (k-h) \sum_{A^h \subseteq A^k} \frac{\varphi_{\ell-1}(A^h)}{\varphi_\ell(A^h)}.
\end{aligned}
$$

Combining this with (2.6.12) and changing ε by $\varepsilon/(k-h)$, we obtain Proposition 2.6.8.

Let $\{A_j^k; j = 1, 2, \dots, q\}$ be a family of decomposable k-vectors in $\mathbf{C}^{n+1}$ which are in general position. Set

$$(2.6.13)\qquad p_k(\ell, h) = \sum_{m=k-h}^{\infty} \binom{n-\ell}{m+1} \binom{\ell+1}{k-m}$$

for $\ell \geq h,\ k \geq h$, where $\binom{k}{h} = 0$ if $h < 0$ or $k < h$ and we set $p_k(\ell, h) = 0$ for $\ell < h$ or $k < h$. Then, we can prove the following:

LEMMA 2.6.14. *For any point $z_0 \in \mathbf{C}$ with $F_\ell(z_0) \neq 0$, there are at most $p_k(\ell, h)$ A_j^k's such that $\varphi_\ell(A^h)(z_0) = 0$ for all $A^h \subseteq A_j^k$.*

PROOF. Take an arbitrary point $z_0 \in \mathbf{C}$ with $F_\ell(z_0) \neq 0$ and set $B := F_\ell(z_0)/|F_\ell(z_0)|$. We define the vector space

$$V(B) := \{X; X \wedge B = 0\},$$

$$V(B)^\perp := \{Z; \langle Z, X\rangle = 0, \text{ for all } X \in V(B)\}.$$

Consider some A_j^k such that $B \vee A^h = 0$ for all $A^h \subseteq A_j^k$ and, for brevity, set $A^k := A_j^k$. For a subspace $V(A) := \{X; X \wedge A = 0\}$, set

$$m := \dim(V(A) \cap V(B)^\perp) - 1.$$

Choose a basis $\{A_0, \dots, A_k\}$ of $V(A)$ such that $\{A_0, \dots, A_m\}$ gives a basis of $V(A) \cap V(B)^\perp$. We write

$$A_{m+1} = X_{m+1} + Y_{m+1}, \dots, A_k = X_k + Y_k \tag{2.6.15}$$

with vectors $X_{m+1}, \dots, X_k \in V(B)^\perp$ and $Y_{m+1}, \dots, Y_k \in V(B)$. Then, $Y_{m+1}, \dots, Y_k$ are linearly independent and so $Y := Y_{m+1} \wedge \dots \wedge Y_k \neq 0$. In fact, if there is some nonzero vector $(c_{m+1}, \dots, c_k)$ such that $c_{m+1}Y_{m+1} + \dots + c_k Y_k = 0$, then we have an absurd conclusion

$$c_{m+1}A_{m+1} + \dots + c_k A_k = c_{m+1}X_{m+1} + \dots + c_k X_k \in V(A) \cap V(B)^\perp.$$

Assume that $m < k - h$. Since $X_j \in V(B)^\perp$ for $j = m+1, \dots, k$, the assumption for A^k implies

$$B \vee Y = B \vee ((A_{m+1} - X_{m+1}) \wedge \dots \wedge (A_k - X_k)) = B \vee (A_{m+1} \wedge \dots \wedge A_k) = 0.$$

This is a contradiction. Therefore, we conclude $m \geq k - h$. By the use of the representation (2.6.15), we see easily

$$A^k \in W := \sum_{m \geq k-h} \bigwedge^{m+1} V(B)^\perp \wedge \bigwedge^{k-m} V(B).$$

On the other hand, W has dimension $p_k(\ell, h)$ because $\dim V(B) = \ell + 1$ and $\dim V(B^\perp) = n - \ell$. Since $\{A_j^k\}$ are in general position, the number of A_j^k's satisfying the condition in Lemma 2.6.14 cannot exceed $p_k(\ell, h)$. This completes the proof of Lemma 2.6.14.

Now, we give the following sum to product estimate for derived curves.

PROPOSITION 2.6.16. *Let $\{A_j^k; j = 1, \ldots, q\}$ be a family of decomposable k-vectors in general position, where $q > \binom{n+1}{k+1}$. Then, there are positive constants a and $c_{\ell h}$ such that*
(2.6.17)

$$\sum_{j=1}^{q} \frac{\psi_{\ell-1,h-1}(A_j^k)}{\psi_{\ell,h}(A_j^k) \prod_{A^{h-1} \subseteq A^h} \log^2(a\varphi_\ell(A^{h-1})/\varphi_\ell(A^h))}$$
$$\geq c_{\ell h} \prod_{j=1}^{q} \left(\frac{\psi_{\ell-1,h-1}(A_j^k)}{\psi_{\ell,h}(A_j^k) \prod_{A^{h-1} \subseteq A^h} \log^2(a\varphi_\ell(A^{h-1})/\varphi_\ell(A^h))} \right)^{1/p_k(\ell,h)}.$$

PROOF. By the same argument as in the proof of Proposition 2.5.2, there is a positive constant δ_{jh} such that

$$\psi_{j,h} \geq \delta_{jh} \tag{2.6.18}$$

for all except at most $p_k(\ell, h)$ j's. Since $\varphi_{\ell-1}(A^{h-1}) \leq \varphi_\ell(A^{h-1})$ as is easily seen with the help of moving frames for f, $\psi_{\ell-1,h-1}(A_j^k)$ is uniformly bounded. Therefore, we can find some $L > 1$, not depending on each point in $\mathbf{C}$, such that

$$\frac{\psi_{\ell-1.h-1}(A_j^k)}{\psi_{\ell,h}(A_j^k) \prod_{A^{h-1} \subseteq A^h} \log^2(a\varphi_\ell(A^{h-1})/\varphi_\ell(A^h))} \leq L$$

for all except at most $p_k(\ell, h)$ j's. We can conclude the proof of Proposition 2.6.16 by the same method as in Proposition 2.5.2, where in this case the quantities $\omega(j)(1 \leq j \leq q), \Phi_{jk}$ and $n - k$ are replaced by $1, \psi_{\ell,h}$ and $p_k(\ell, h)$ respectively.

Chapter 3

The classical defect relations for holomorphic curves

§3.1 The first main theorem for holomorphic curves

We begin by explaining the counting functions of divisors.

For $0 \leq R \leq \infty$ set $\Delta_R := \{z \in \mathbf{C}; |z| < R\}$, where Δ_∞ means the complex plane $\mathbf{C}$. We also set $\Delta_{s,R} := \Delta_R - \Delta_s$ for $0 \leq s < R$. In particular, we have $\Delta_{0,R} = \Delta_R$. Consider a divisor ν on $\Delta_{s,R}$. For brevity, we assume that $\nu(0) = 0$ for the particular case $s = 0$. For $s < t < R$ we set

$$n(t,\nu) := \sum_{s \leq |z| \leq t} \nu(z),$$

which is finite because only finitely many terms are nonzero.

DEFINITION 3.1.1. We define the *counting function* of ν by

$$N(r,\nu) = \int_s^r \frac{n(t,\nu)}{t} dt \qquad (s \leq r < R).$$

We can show the following:

$$N(r,\nu) = \sum_{a \in \Delta_{s,R}} \nu(a) \log^+ \frac{r}{|a|}, \tag{3.1.2}$$

where $\log^+ x = \max(\log x, 0)$.

To see this, for each point a we set

$$\nu^a(z) = \begin{cases} 1 & z = a, \\ 0 & z \neq a. \end{cases}$$

Then we can write

$$\nu = \nu(a_1)\nu^{a_1} + \cdots + \nu(a_k)\nu^{a_k}$$

when $\mathrm{Supp}(\nu)\cap\{z; s \leq |z| \leq r\} = \{a_1, \ldots, a_k\}$. Since both sides of (3.1.2) are linear in ν, it suffices to show (3.1.2) in the case that $\nu = \nu^a$ for an arbitrarily given a. In this case, we see easily

$$n(t, \nu^a) = \begin{cases} 0 & s \leq t < |a|, \\ 1 & |a| \leq t \end{cases}$$

and so $N(r, \nu^a) = 0$ for $r < |a|$. Therefore,

$$N(r, \nu^a) = \int_{|a|}^{r} \frac{1}{t} dt = \log \frac{r}{|a|}$$

for $r \geq |a|$, which completes the proof of (3.1.2).

We give here the following important formula related to counting functions.

PROPOSITION 3.1.3. *Let u be a function with mild singularities on $\Delta_{s,R}$. Assume that $\nu_u(0) = 0$ for the particular case $s = 0$. For $s < r < R$ it holds that*

$$\begin{aligned} \int_s^r \frac{dt}{t} \int_{\Delta_{s,t}} & dd^c \log |u|^2 + N(r, \nu_u) \\ & = \frac{1}{2\pi} \int_0^{2\pi} \log |u(re^{i\theta})| d\theta - \frac{1}{2\pi} \int_0^{2\pi} \log |u(se^{i\theta})| d\theta \\ & \qquad - A(s) \log r + B(s), \end{aligned} \tag{3.1.4}$$

where

$$A(s) = \int_{|z|=s} d^c \log |u|^2, \ B(s) = A(s) \log s$$

for the case $s > 0$ and $A(0) = B(0) = 0$.

PROOF. To see this, we may replace the given function u by $|u|$, and so u may be assumed to be nonnegative and real-valued outside a discrete set.

We first show the following Stokes theorem for functions with mild singularities.

(3.1.5) *Let D be a relatively compact domain with smooth boundary and u a nonnegative function with mild singularities on a neighborhood of $\bar{D}$. If $\nu_u \equiv 0$ and u is smooth on ∂D, then we have*

$$\int_D dd^c \log |u|^2 = \int_{\partial D} d^c \log |u|^2.$$

To see this, we may assume that u is a positive C^∞ function on $D - \{a_1, \dots, a_k\}$. We take a holomorphic local coordinates z_i around each a_i with $z_i(a_i) = 0$. By Stokes theorem for smooth functions, we have

$$\begin{aligned}\int_D dd^c \log|u|^2 &= \lim_{\varepsilon\to 0}\int_{D-\cup_i\{|z_i|\le\varepsilon\}} dd^c\log|u|^2\\ &= \int_{\partial D} d^c\log|u|^2 - \lim_{\varepsilon\to 0}\sum_i\int_{\{|z_i|=\varepsilon\}} d^c\log|u|^2.\end{aligned}$$

We have only to show that the last term in the above is equal to zero. The problem is local. It suffices to prove that

$$I(\varepsilon) := \int_{\{|z|=\varepsilon\}} d^c\log|u|^2$$

converges to zero as $\varepsilon \to 0$ for a function

$$u := |\log(|z|^m v(z))|,$$

where m is an integer and v is a positive C^∞ function near the origin. In this situation, using the formula

$$\text{(3.1.6)}\qquad d = \frac{\partial}{\partial r}dr + \frac{\partial}{\partial\theta}d\theta, \qquad d^c = \frac{1}{4\pi}r\frac{\partial}{\partial r}d\theta - \frac{1}{4\pi r}\frac{\partial}{\partial\theta}dr$$

for the polar coordinate $z = re^{i\theta}$, we can easily check that

$$I(\varepsilon) = O\left(\frac{1}{|\log\varepsilon|}\right),$$

which converges to zero as $\varepsilon \to 0$. Thus, we have the assertion (3.1.5).

The proof of (3.1.4) is divided into four steps.

1°. Firstly, we consider the case where u is a positive C^∞ function on $\{r_1 \le |z| \le r_2\}$ for some r_1, r_2 with $s < r_1 < r_2 < R$ and $\nu_u \equiv 0$ on Δ_{r_1,r_2}. By (3.1.5), we have

$$\begin{aligned}\int_{r_1}^{r_2}\frac{dt}{t}\int_{\Delta_{s,t}} dd^c\log|u|^2 &= \int_{r_1}^{r_2}\frac{dt}{t}\left(\int_{|z|=t} d^c\log|u|^2 - A(s)\right)\\ &= \int_{r_1}^{r_2}\frac{dt}{t}\left(\int_{|z|=t} d^c\log|u|^2\right) - A(s)\log\frac{r_2}{r_1}.\end{aligned}$$

Since $dr = 0$ on $\{z; |z| = t\}$, by the use of (3.1.6) we get

$$\begin{aligned}\int_{r_1}^{r_2} \frac{dt}{t} \int_{|z|=t} & d^c \log |u|^2 \\ &= \frac{1}{2\pi} \int_{r_1}^{r_2} \frac{dt}{t} \int_0^{2\pi} t \frac{\partial}{\partial t} \log |u(te^{i\theta})| d\theta \\ &= \frac{1}{2\pi} \int_0^{2\pi} d\theta \frac{\partial}{\partial t} \int_{r_1}^{r_2} \log |u(te^{i\theta})| dt \\ &= \frac{1}{2\pi} \int_0^{2\pi} \log |u(r_2 e^{i\theta})| d\theta - \frac{1}{2\pi} \int_0^{2\pi} \log |u(r_1 e^{i\theta})| d\theta.\end{aligned}$$

Therefore, in this case, we obtain

$$\begin{aligned}\int_{r_1}^{r_2} \frac{dt}{t} \int_{\Delta_{s,t}} dd^c \log |u|^2 &= \frac{1}{2\pi} \int_0^{2\pi} \log |u(r_2 e^{i\theta})| d\theta \\ &- \frac{1}{2\pi} \int_0^{2\pi} \log |u(r_1 e^{i\theta})| d\theta - A(s) \log \frac{r_2}{r_1}.\end{aligned} \tag{3.1.7}$$

In particular, if u is a positive C^∞-function on a neighborhood of $\{z; s \leq |z| \leq r\}$, we obtain (3.1.4) by applying (3.1.7) to $r_1 := s$ and $r_2 := r$.

2°. We consider next the particular case where $u = |z - a|^\sigma$ for some real number σ and a point a in $\Delta_{s,R}$. In this case, the first term of the left hand side of (3.1.4) vanishes because $\log |u|$ is harmonic on $\Delta_{s,R}$ except at one point a, and $N(r, \nu_u) = \sigma \log^+ |r/a|$ by (3.1.2). We set

$$J(r) := \frac{1}{2\pi} \int_0^{2\pi} \log |re^{i\theta} - a| d\theta$$

If $r < a$, then we find $J(r) = \log |a|$ by applying the mean value theorem for a harmonic function $\log |z - a|$ on $\Delta_{|a|}$. Moreover, if $r > a$, then

$$\begin{aligned}J(r) &= \frac{1}{2\pi} \int_0^{2\pi} \log \left| \frac{r}{a} - e^{i\theta} \right| d\theta + \log |a| \\ &= \log \frac{r}{|a|} + \log |a| = \log r,\end{aligned}$$

because $\log \left| \frac{r}{a} - z \right|$ is harmonic on $\Delta_{r/|a|}$. Therefore, by monotonicity or continuity of $J(r)$, we have

$$J(r) = \log^+ \frac{r}{|a|} + \log |a|$$

for all positive numbers r. Since $A(s) = 0$ in this case, the right hand side of (3.1.4) is equal to $\sigma(J(r) - \log|a|) = N(r, \nu_u)$. Therefore, (3.1.4) holds in this particular case.

3°. We next consider the case where $\nu_u \equiv 0$. In this case, u is positive and of class C^∞ outside a set $\cup_{i=1}^k \{z; |z| = s_i\}$, where $s_0 := s \leq s_1 < \cdots < s_k \leq s_{k+1} := r$. Moreover, both sides of (3.1.7) are continuous in r_1 and r_2, because for a nonzero holomorphic function $g(z)$ and positive C^∞ function v the singularities of $\log|\log|g(z)v(z)||$ are very mild and the function $dd^c \log|\log|g(z)v(z)||$ is locally integrable. Therefore, we have (3.1.7) if we set $r_1 := s_i$ and $r_2 := s_{i+1}$ for $i = 0, \ldots, k$. By summing up these identities, we have (3.1.4) easily in this case too.

4°. For the other cases, u can be written as the finite product of functions of the types considered in 1°, 2° and 3°. With the help of linearity of counting functions in ν, we conclude the identity (3.1.4).

For convenience's sake, we set

$$\eta_s(r) = \begin{cases} O(\log r) & \text{for } s > 0, \\ 0 & \text{for } s = 0. \end{cases}$$

Since $dd^c \log|g|^2 = 0$ for a nonzero meromorphic function g, we get the following Jensen's formula as a consequence of Proposition 3.1.3.

COROLLARY 3.1.8. *Let g be a nonzero meromorphic function on (some open neighborhood of) $\Delta_{s,R}$. For the particular case $s = 0$ assume that g has neither a zero nor a pole at the origin. Then, it holds that*

$$N(r, \nu_g) = \frac{1}{2\pi}\int_0^{2\pi} \log|g(re^{i\theta})|d\theta - \frac{1}{2\pi}\int_0^{2\pi} \log|g(se^{i\theta})|d\theta + \eta_s(r).$$

Now, take a nondegenerate holomorphic curve f in $P^n(\mathbf{C})$ defined on $\Delta_{s,R}$. We consider the k-th derived curve f^k of f for each $k = 0, \ldots, n-1$.

DEFINITION 3.1.9. The *order function* of f^k is defined by

$$T_f^k(r) = \int_s^r \frac{dt}{t} \int_{\Delta_{s,t}} \Omega_k.$$

In particular, define $T_f(r) := T_f^0(r)$. For convenience's sake, we set $T_f^{-1}(r) \equiv 0$.

Choosing a reduced representation $f = (f_0 : \cdots : f_n)$, we set $F := (f_0, \ldots, f_n)$ and $F_k := F \wedge F' \wedge \ldots \wedge F^{(k)}$, and define

$$\nu_k := \nu_{|F_k|} \tag{3.1.10}$$

and

$$N_f^k(r) := N(r, \nu_k).$$

Since $f_0, \ldots, f_n$ have no common zero, we have $N_f^0(r) \equiv 0$.

By substituting $u := |F_k|$ into (3.1.4) we can conclude the following:

PROPOSITION 3.1.11. *In the above situation, it holds that*

$$\begin{aligned} &T_f^k(r) + N_f^k(r) \\ &\quad = \frac{1}{2\pi}\int_0^{2\pi} \log|F_k(re^{i\theta})|d\theta - \frac{1}{2\pi}\int_0^{2\pi} \log|F_k(se^{i\theta})|d\theta + \eta_s(r). \end{aligned}$$

For the particular case $k = 0$, we have

COROLLARY 3.1.12. *For a holomorphic curve $f : \Delta_{s,R} \to P^n(\mathbf{C})$, we have*

$$T_f(r) = \frac{1}{2\pi}\int_0^{2\pi} \log|F(re^{i\theta})|d\theta - \frac{1}{2\pi}\int_0^{2\pi} \log|F(se^{i\theta})|d\theta + \eta_s(r),$$

where $F := (f_0, \ldots, f_n)$ for a reduced representation $f = (f_0 : \cdots : f_n)$.

Take a hyperplane

$$H : \langle W, A \rangle = 0$$

in $P^{N_k}(\mathbf{C})$ not including the image of f^k, where $N_k := \binom{n+1}{k+1} - 1$, A is a unit vector in $\bigwedge^{k+1} \mathbf{C}^{n+1}$ and W denotes homogeneous coordinates on $P^{N_k}(\mathbf{C})$. The notation $\nu(f^k, H)$ denotes the pull-back of H via f^k considered as a divisor. Take a nonzero holomorphic function h such that $\nu_h = \nu_{|F_k|}$. Then F_k/h gives a reduced representation of f^k. Therefore, we have

$$\nu(f^k, H) = \nu_{\langle F_k, A\rangle} - \nu_{|F_k|}.$$

DEFINITION 3.1.13. Let m be a nonnegative integer or $m = \infty$. We define the *counting function* of H for f^k (truncated by m) by

$$N_f^k(r, H)^{[m]} := N(r, \min(\nu(f^k, H), m)).$$

For brevity, we set $N_f^k(r,H) := N_f^k(r,H)^{[\infty]}$, $N_f(r,H)^{[m]} = N_f^0(r,H)^{[m]}$ and $N_f(r,H) := N_f^0(r,H)^{[\infty]}$.

We define

$$\psi(f^k,H) := \frac{|F_k|}{|\langle F_k, A\rangle|} \quad (\geq 1).$$

DEFINITION 3.1.14. We define the *proximity function* of H for f^k by

$$m_f^k(r,H) = \frac{1}{2\pi}\int_0^{2\pi} \log \psi(f^k,H)(re^{i\theta})d\theta (\geq 0).$$

For brevity, we set $m_f(r,H) = m_f^0(r,H)$.

Now, we give the first main theorem for value distribution theory of the derived curves of holomorphic curves.

THEOREM 3.1.15. *Let f be a nondegenerate holomorphic curve in $P^n(\mathbf{C})$ defined on $\Delta_{s,R}$, H a hyperplane in $P^{N_k}(\mathbf{C})$ and f^k the derived curves of f. For the particular case $s=0$ we assume that $f^k(0) \notin H$ and $\nu_k(0) = 0$. Then, it holds that*

$$T_f^k(r) = N_f^k(r,H) + m_f^k(r,H) + \eta_s(r).$$

PROOF. Apply Proposition 3.1.3 to the function $u := |F_k/h|$, where h is a nonzero holomorphic function with $\nu_h = \nu_{|F_k|}$. Since F_k/h gives a reduced representation of f^k and

$$|F_k| = |\langle F_k, A\rangle|\, \psi(f^k,H),$$

we see

$$\begin{aligned} T_f^k(r) &= \frac{1}{2\pi}\int_0^{2\pi} \log|F_k/h|(re^{i\theta})d\theta - \frac{1}{2\pi}\int_0^{2\pi} \log|F_k/h|(se^{i\theta})d\theta + \eta_s(r) \\ &= \frac{1}{2\pi}\int_0^{2\pi} \log\left|\frac{\langle F_k(re^{i\theta}), A\rangle}{h(re^{i\theta})}\right| d\theta \\ &\qquad - \frac{1}{2\pi}\int_0^{2\pi} \log\left|\frac{\langle F_k(se^{i\theta}), A\rangle}{h(se^{i\theta})}\right| d\theta \\ &\qquad + m_f^k(r,H) - m_f^k(s,H) + \eta_s(r) \\ &= N_f^k(r,H) + m_f^k(r,H) - m_f^k(s,H) + \eta_s(r) \end{aligned}$$

with the help of Corollary 3.1.8. This proves Theorem 3.1.15.

As the particular case $k=0$ of Theorem 3.1.15 we have the following first main theorem for holomorphic curves.

COROLLARY 3.1.16. *Let $f : \Delta_{s,R} \to P^n(\mathbf{C})$ be a nondegenerate holomorphic curve and H a hyperplane in $P^n(\mathbf{C})$. Assume that $f(0) \notin H$ for the particular case $s = 0$. Then, it holds that*

$$T_f(r) = N_f(r, H) + m_f(r, H) + \eta_s(r).$$

§3.2 The second main theorem for holomorphic curves

We consider a nondegenerate holomorphic curve f in $P^n(\mathbf{C})$ defined on an open neighborhood of $\Delta_{s,\infty} := \{z; s \leq |z| < +\infty\}$ and study the order functions $T_f^k(r)$ of the derived curves f^k of f. Set

$$\Omega_k = h_k dd^c |z|^2.$$

For brevity, we assume $h_k(0) \neq 0$ for all k in the case $s = 0$.

By definition,

$$T_f^k(r) = \int_s^r \frac{dt}{t} \int_{\Delta_{s,t}} h_k dd^c |z|^2.$$

We define the *divisor of ramification* of the k-th derived curves of f by

$$\mu_k := \frac{1}{2} \nu_{h_k} (\geq 0). \tag{3.2.1}$$

THEOREM 3.2.2. *For each $k = 0, 1, \ldots, n-1$ it holds that*

$$\begin{aligned} T_f^{k-1}(r) &- 2T_f^k(r) + T_f^{k+1}(r) + N(r, \mu_k) \\ &= \frac{1}{4\pi} \int_0^{2\pi} \log h_k(re^{i\theta}) d\theta - \frac{1}{4\pi} \int_0^{2\pi} \log h_k(se^{i\theta}) d\theta + \eta_s(r), \end{aligned}$$

where $\eta_s(r) = O(\log r)$ for $s > 0$ and $\eta_s(r) = 0$ for $s = 0$ as in the previous section.

PROOF. By virtue of Proposition 2.2.11, we have

$$dd^c \log h_k = \Omega_{k-1} - 2\Omega_k + \Omega_{k+1}.$$

Integrating this identity twice, we can easily conclude Theorem 3.2.2 by substituting $u = h_k^{1/2}$ in Proposition 3.1.3.

Now, for a nonnegative real-valued locally summable function $h(\not\equiv 0)$ on $\Delta_{s,\infty}$, we define the *order function* of h by

$$T^h(r) := \int_s^r \frac{dt}{t} \int_{\Delta_{s,t}} h \, dd^c|z|^2. \tag{3.2.3}$$

PROPOSITION 3.2.4. *There are positive numbers C_0 and C_1, not depending on r, and a subset E of $[s,+\infty)$ with $\int_E (1/r)dr < +\infty$ such that*

$$\int_0^{2\pi} \log h(re^{i\theta})d\theta \le C_0 \log T^h(r) + C_1 \qquad (r \notin E).$$

For the proof, we use the following lemma of E. Borel.

LEMMA 3.2.5. *Let $u(r), v(r)$ be positive increasing differentiable functions on $[s,+\infty)$ such that $u'(r)$ is continuous and $v'(r)$ is piecewise continuous, and let $a(t)$ be a positive increasing function on $[t_0,+\infty)$, where $u([s,+\infty)) \subset [t_0,+\infty)$. Set*

$$E := \{r \in [s,+\infty); u'(r) \ge v'(r)a(u(r))\}.$$

Then,

$$\int_E v'(r)dr \le \int_{t_0}^{+\infty} \frac{dt}{a(t)}.$$

PROOF. This is obvious from the inequalities

$$\int_E v'(r)dr \le \int_E \frac{u'(r)}{a(u(r))}dr \le \int_{t_0}^{+\infty} \frac{dt}{a(t)}.$$

PROOF OF PROPOSITION 3.2.4. By definition, we have

$$T^h(r) = \int_s^r \frac{dt}{t}\frac{1}{\pi}\int_s^t \rho d\rho \int_0^{2\pi} h(\rho e^{i\theta})d\theta,$$

because $dd^c|z|^2 = \dfrac{1}{\pi}\rho d\rho \wedge d\theta$ for $z = \rho e^{i\theta}$. This implies that

$$\frac{dT^h}{d\log r} = r\frac{dT^h}{dr} = \frac{1}{\pi}\int_s^r \rho d\rho \int_0^{2\pi} h(\rho e^{i\theta})d\theta \ge 0.$$

Hence, $T^h(r)$ is an increasing function in log r and so in r. Moreover, it holds that

$$\frac{1}{r^2}\frac{d^2T^h}{(d\log r)^2} = \frac{1}{r}\frac{d}{dr}\left(\frac{dT^h}{d\log r}\right) = \frac{1}{\pi}\int_0^{2\pi} h(re^{i\theta})d\theta.$$

Using the concavity of the logarithm, we obtain

$$\begin{aligned}\frac{1}{2\pi}\int_0^{2\pi}\log h(re^{i\theta})d\theta &\le \log\left(\frac{1}{2\pi}\int_0^{2\pi} h(re^{i\theta})d\theta\right)\\ &= \log\left(\frac{1}{2r^2}\frac{d^2T^h}{(d\log r)^2}\right).\end{aligned}$$

We now apply Lemma 3.2.5 to the functions $u(r) = T^h(r),\ v(r) = \log r$ and $a(t) = t^{1+\varepsilon}$ for a sufficiently small ε to show that

$$\frac{dT^h(r)}{dr} \le \frac{1}{r}T^h(r)^{1+\varepsilon}$$

excluding a set E_1 with

$$\int_{E_1}\frac{dr}{r} \le \int_{t_0}^{+\infty}\frac{dt}{t^{1+\varepsilon}} < +\infty.$$

Moreover, by applying Lemma 3.2.5 once more to the function

$$u(r) = \frac{dT^h(r)}{d\log r}$$

and the above functions $v(r), a(t)$, we get

$$\frac{d^2T^h(r)}{(d\log r)^2} \le \left(\frac{dT^h(r)}{d\log r}\right)^{1+\varepsilon} \le T^h(r)^{(1+\varepsilon)^2}.$$

excluding the union E of E_1 and another set E_2 with $\int_{E_2}(1/r)dr < +\infty$. Therefore, we conclude

$$\int_0^{2\pi}\log h(re^{i\theta})d\theta \le C_0\log T^h(r) + C_1$$

excluding E for positive constants C_0 and C_1. This gives Proposition 3.2.4.

COROLLARY 3.2.6. *Let h be a nonnegative function $h(\not\equiv 0)$ with mild singularities on $\Delta_{s,\infty}$. If $\nu_h \geq 0$, then there are some positive constants C_0, C_1 and a set E of finite logarithmic measure such that*

$$\int_s^r \frac{dt}{t} \int_{\Delta_{s,t}} dd^c \log h \leq C_0 \log T^h(r) + C_1 + \eta_s(r) \qquad (r \notin E).$$

PROOF. This is an immediate consequence of Propositions 3.1.3 and 3.2.4.

Now, we set

$$T(r) := \max_{0 \leq k \leq n-1} T_f^k(r)$$

and denote by $\alpha(r)$ some function in r such that

$$\alpha(r) = \begin{cases} O(\log T(r)) + O(\log r) & \text{for the case } s > 0, \\ O(\log T(r)) + O(1) & \text{for the case } s = 0 \end{cases}$$

excluding a set E of finite logarithmic measure. As a result of Theorem 3.2.2 and Proposition 3.2.4, we see

$$T_f^{k-1}(r) - 2T_f^k(r) + T^{k+1}(r) \leq \alpha(r). \tag{3.2.7}$$

Moreover, we can prove the following:

PROPOSITION 3.2.8. *It holds that*

(i) $\dfrac{T_f^\ell(r)}{\ell+1} \leq \dfrac{T_f^k(r)}{k+1} + \alpha(r)$ *for* $\ell \geq k$,

(ii) $\dfrac{T_f^\ell(r)}{n-\ell} \leq \dfrac{T_f^k(r)}{n-k} + \alpha(r)$ *for* $\ell \leq k$.

PROOF. We first show that

$$\frac{T_f^k(r)}{k+1} \leq \frac{T_f^{k-1}(r)}{k} + \alpha(r). \tag{3.2.9}$$

This is proved by induction on k. For the case $k = 1$, (3.2.9) is obvious by (3.2.7). Assume that it is true for the case $\leq k$. Then, by the use of (3.2.7) we see

$$\begin{aligned}(k+1)T_f^{k+1} &\leq 2(k+1)T_f^k - (k+1)T_f^{k-1} + \alpha(r) \\ &\leq (k+2)T_f^k + \alpha(r).\end{aligned}$$

In the similar manner, by downward induction on k starting from $k = n-1$ in this case, we can prove

$$\frac{T_f^{k-1}}{n-(k-1)} \leq \frac{T_f^k(r)}{n-k} + \alpha(r). \tag{3.2.10}$$

The identities (i) and (ii) of Proposition 3.2.8 for the case $\ell = k$ are both trivial. For the general cases, (i) is also easily shown by upward induction on ℓ with the use of (3.2.9). The identity (ii) is similarly proved with the use of (3.2.10) by downward induction on ℓ.

COROLLARY 3.2.11. *For all k and ℓ it holds that*

$$T_f^k(r) \leq O(T_f^\ell(r)) + O(\log r)$$

for all r not contained in a set E of finite logarithmic measure, where the last term $O(\log r)$ can be replaced by $O(1)$ for the case $s = 0$.

PROOF. This is obvious for the case where $T(r)$ is bounded. Assume that $\lim_{r\to\infty} T(r) = +\infty$. By Proposition 3.2.8 we see easily that

$$T(r) \leq O(T_f^\ell(r)) + \alpha(r) = O(T_f^\ell(r)) + O(\log T(r)) + O(\log r)$$

for all ℓ and for all r not contained in a set E with finite logarithmic measure. Therefore,

$$T_f^k \leq T(r) \leq \frac{O(T_f^\ell(r)) + O(\log r)}{1 - O\left(\dfrac{\log T(r)}{T(r)}\right)} \qquad (r \notin E).$$

Since $\displaystyle\lim_{r\to\infty} \frac{\log T(r)}{T(r)} = 0$, we can conclude Corollary 3.2.11. For the case $s = 0$, we can replace $O(\log r)$ by $O(1)$ in the above argument.

Now, we give the second main theorem for a holomorphic curve in $P^n(\mathbf{C})$ defined on $\Delta_{s,\infty}$.

THEOREM 3.2.12. *Let f be a nondegenerate holomorphic curve in $P^n(\mathbf{C})$, let $H_1, H_2, \ldots, H_q$ be hyperplanes in $P^n(\mathbf{C})$ located in N-subgeneral position and let $\omega(j)$ and θ be Nochka weights and a Nochka constant respectively for these hyperplanes, where $q \geq 2N - n + 1$. Then, for every $\varepsilon > 0$ there exists a set E with $\int_E (1/r)dr < +\infty$ such that, for all $r \notin E$,*

$$\theta(q - 2N + n - 1 - \varepsilon)T_f(r) \leq \sum_{j=1}^{q} \omega(j) N_f(r, H_j)^{[n]} + O(\log T_f(r)) + \eta_s(r),$$

where $\eta_s(r) = O(\log r)$ for $s > 0$ and $\eta_s(r) = O(1)$ for $s = 0$.

For the proof, we need the following:

LEMMA 3.2.13. *Let f be a nondegenerate holomorphic curve in $P^n(\mathbf{C})$ with a reduced representation $f = (f_0 : \cdots : f_n)$ defined on a domain D in $\mathbf{C}$. Consider q hyperplanes*

$$H_j : \langle W, A_j \rangle = 0 \qquad (1 \leq j \leq q)$$

in $P^n(\mathbf{C})$ in N-subgeneral position and take Nochka weights $\omega(1), \dots, \omega(q)$ for these hyperplanes, where $q > 2N - n + 1$. Set $F(H_j) = \langle F, A_j \rangle$ for $F = (f_0, \dots, f_n)$ and

$$\varphi = \frac{|W(f_0, \dots, f_n)|}{|F(H_1)|^{\omega(1)} \cdots |F(H_q)|^{\omega(q)}}.$$

Then,

$$\nu_\varphi + \sum_{j=1}^{q} \omega(j) \min(\nu(f, H_j), n) \geq 0.$$

PROOF. For brevity, we set $W := W(f_0, \dots, f_n)$. It suffices to show that

$$\nu_W \geq \sum_{j=1}^{q} \omega(j)(\nu(f, H_j) - n)^+, \tag{3.2.14}$$

where x^+ means $\max(x, 0)$ for a real number x. In fact, since

$$\min(\nu(f, H_j), n) + (\nu(f, H_j) - n)^+ = \nu(f, H_j),$$

we can conclude from (3.2.14)

$$\begin{aligned}
\nu_\varphi + &\sum_{j=1}^{q} \omega(j) \min(\nu(f, H_j), n) \\
&= \nu_W - \sum_{j=1}^{q} \omega(j)\nu(f, H_j) + \sum_{j=1}^{q} \omega(j) \min(\nu(f, H_j), n) \\
&\geq \sum_{j=1}^{q} \omega(j)((\nu(f, H_j) - n)^+ + \min(\nu(f, H_j), n) - \nu(f, H_j)) \\
&= 0.
\end{aligned}$$

To show (3.2.14), take an arbitrary point ζ. We set

$$S = \{j \in Q; \nu(f, H_j)(\zeta) \geq n+1\},$$

where $Q = \{1, 2, \dots, q\}$. We may assume that $S \neq \phi$. Then, $\#S \leq N$. Otherwise, by the assumption of N-nondegeneracy there are $n+1$ distinct indices $j_0, \dots, j_n$ in S such that $A_{j_0}, \dots, A_{j_n}$ are linearly independent and so $f_0, \dots, f_n$ are represented as linear combinations of $F(H_{j_0}), \dots, F(H_{j_n})$. Then, $f_0, \dots, f_n$ vanish at ζ, which is a contradiction.

We now consider the sets $S_\tau (0 \leq \tau \leq t)$ such that

$$S_0 := \phi \neq S_1 \subset S_2 \subset \cdots \subset S_t := S$$

and $\nu(f, H_j)(\zeta)$ equals some constant m_τ for each $j \in S_\tau - S_{\tau-1}$, where $m_1 > m_2 > \cdots > m_t$. We denote by $V(R)$ the vector subspace of $\mathbf{C}^{n+1}$ generated by $\{A_j; j \in R\}$ for a subset R of Q. Then,

$$V(S_1) \subseteq V(S_2) \subseteq \cdots \subseteq V(S_t).$$

For each τ take a subset T_τ of S_τ such that $T_{\tau-1} \subset T_\tau$ and $\{A_j; j \in T_\tau\}$ gives a basis of $V(S_\tau)$. We then have $\#(T_\tau - T_{\tau-1}) = d(S_\tau) - d(S_{\tau-1})$, where $d(S_\tau) := \dim V(S_\tau)$ as in §2.4. For brevity, we set $m_\tau^* := m_\tau - n$. By the use of Theorem 2.4.11, (iv), we obtain

$$\begin{aligned}
\sum_{j=1}^{q} \omega(j)(\nu(f, H_j) - n)^+ &= \sum_{j \in S} \omega(j)(\nu(f, H_j) - n) \\
&= \sum_{\tau=1}^{t} \sum_{j \in S_\tau - S_{\tau-1}} \omega(j) m_\tau^* \\
&= \sum_{\tau=1}^{t} \sum_{j \in S_\tau - S_{\tau-1}} \omega(j) \left(\sum_{\sigma=\tau}^{t-1} (m_\sigma^* - m_{\sigma+1}^*) + m_t^* \right) \\
&= (m_1^* - m_2^*) \sum_{j \in S_1} \omega(j) + (m_2^* - m_3^*) \sum_{j \in S_2} \omega(j) + \cdots + m_t^* \sum_{j \in S_t} \omega(j) \\
&\leq d(S_1)(m_1^* - m_2^*) + d(S_2)(m_2^* - m_3^*) + \cdots + d(S_t) m_t^* \\
&= d(S_1) m_1^* + (d(S_2) - d(S_1)) m_2^* + \cdots + (d(S_t) - d(S_{t-1})) m_t^* \\
&= \#T_1 m_1^* + \#(T_2 - T_1) m_2^* + \cdots + \#(T_t - T_{t-1}) m_t^*.
\end{aligned}$$

Set $T_t = \{j_0, \dots, j_k\}$, where $k \le n+1$. Since $A_{j_0}, \dots, A_{j_k}$ are linearly independent, after a suitable nonsingular linear transformation of homogeneous coordinates we may assume that $f_0 = F(H_{j_0}), \dots, f_k = F(H_{j_k})$. Then, with the Laplace expansion theorem for the determinant, the Wronskian is expanded as the sum of the products of some minors of degree $n-k$ and some minors of degree $k+1$ whose components consist of the $\le n$-th derivatives of the functions $F(H_{j_0}), \dots, F(H_{j_k})$. This implies that

$$\nu_W(\zeta) \ge \sum_{j \in T_t} (\nu(f, H_j)(\zeta) - n).$$

Since $\nu(f, H_j)(\zeta) = m_\tau$ for every $j \in T_\tau - T_{\tau-1}$, this quantity coincides with the last term of the above inequalities. This completes the proof of Lemma 3.2.13.

PROOF OF THEOREM 3.2.12. Choose holomorphic functions g_k with $\nu_{g_k} = \nu_{|F_k|}$ and set $\tilde{F}_k = F_k/g_k$ for each $k = 0, 1, \dots, n$. For a suitably chosen positive numbers ε and a, if we set

$$\hat{h} = \frac{\prod_{k=0}^{n-1} |\tilde{F}_k|^{2\varepsilon}}{\prod_{k=0}^{n-1} \prod_{j=1}^{q} \log^{2\omega(j)}(a/\varphi_k(H_j))},$$

then the function h^* defined by

$$dd^c \log \hat{h} = h^* dd^c |z|^2$$

satisfies the inequality

$$h^{*n(n+1)/2} \ge C \frac{|F_0|^{2\theta(q-2N+n-1)} |F_n|^2 \hat{h}}{|\tilde{F}_0|^{2\varepsilon} \cdots |\tilde{F}_{n-1}|^{2\varepsilon} \prod_{j=1}^{q} |F(H_j)|^{2\omega(j)}}$$

for a positive constant C by Corollary 2.5.5. Apply the operator $\frac{1}{4\pi}\int_0^{2\pi} \log$ to both sides of this inequality. Then, for the function

$$\varphi := \frac{|F_n|}{\prod_{j=1}^{q} |F(H_j)|^{\omega(j)}},$$

we obtain

$$\begin{aligned} &\frac{n(n+1)}{2} \frac{1}{4\pi} \int_0^{2\pi} \log h^*(re^{i\theta}) d\theta + \eta_s(r) \\ &\ge \frac{1}{4\pi} \int_0^{2\pi} \log \left(\frac{|F_0|^{2\theta(q-2N+n-1)} |\varphi|^2 \hat{h}}{|\tilde{F}_0|^{2\varepsilon} \cdots |\tilde{F}_{n-1}|^{2\varepsilon}} \right) d\theta + \eta_s(r) \qquad (3.2.15) \\ &\ge \theta(q - 2N + n - 1) T_f(r) + N(r, \nu_\varphi) \\ &\quad - \varepsilon(T_f^0(r) + \cdots + T_f^{n-1}(r)) + \frac{1}{4\pi} \int_0^{2\pi} \log \hat{h}(re^{i\theta}) d\theta \end{aligned}$$

with the use of Proposition 3.1.11, Corollaries 3.1.8 and 3.1.12. To study the first term of the left side of (3.2.15), apply Proposition 3.2.4, Corollary 3.2.6 and Proposition 3.1.3 and we see

$$\begin{aligned}\int_0^{2\pi} \log h^*(re^{i\theta})d\theta &\leq C_0 \log\left(\int_s^r \frac{dt}{t}\int_{\Delta_{s,t}} h^* dd^c|z|^2\right) + C_1\\ &= C_0 \log\left(\int_s^r \frac{dt}{t}\int_{\Delta_{s,t}} dd^c \log \hat{h}\right) + C_1\\ &\leq C_0 \log\left(\int_0^{2\pi} \log \hat{h}(re^{i\theta})d\theta + \eta_s(r)\right) + C_1\\ &\leq C_0 \log\left(\int_0^{2\pi} \log \hat{h}(re^{i\theta})d\theta\right) + \eta_s(r)\end{aligned}$$

for a positive constant C_0, C_1 and each $r \in [s, +\infty)$ except possibly on a set with finite logarithmic measure. On the other hand, by Lemma 3.2.13 we have

$$N(r, \nu_\varphi) \geq -\sum_{j=1}^{q} \omega(j) N_f(r, H_j)^{[n]}.$$

Moreover, $\log\left(\int_0^{2\pi} \log \hat{h}(re^{i\theta})d\theta\right) - \int_0^{2\pi} \log \hat{h}(re^{i\theta})d\theta$ is bounded from above, and by Corollary 3.2.11 each $\varepsilon T_f^k(r)$ is bounded by $\varepsilon T_f(r) + \eta_s(r)$ outside a set with finite logarithmic measure after a suitable change of ε. From these facts we can easily conclude that

$$\theta(q - 2N + n - 1 - \varepsilon)T_f(r) \leq \sum_{j=1}^{q} \omega(j) N_f(r, H)^{[n]} + O(\log T_f(r)) + \eta_s(r)$$

outside a set of finite logarithmic measure. This gives Theorem 3.2.12.

§3.3 Defect relations for holomorphic curves

As in the previous section, we consider a holomorphic curve $f : \Delta_{s,\infty} \to P^n(\mathbf{C})$ with a reduced representation $f = (f_0 : \cdots : f_n)$. We first give some elementary properties of the order function of f.

PROPOSITION 3.3.1. *If f is not a constant, then*

$$\lim_{r \to +\infty} T_f(r) = +\infty.$$

PROOF. The nonnegative form $\Omega_0 = dd^c \log(|f_0|^2 + \cdots + |f_n|^2)$ does not vanish identically. Because, otherwise, $F_1 \equiv 0$ by Proposition 2.2.11, which implies that f is a constant. Therefore, there is some t_0 with $s < t_0$ such that

$$K := \int_{\Delta_{s,t_0}} \Omega_0 > 0.$$

It follows that, for $r > t_0$,

$$\begin{aligned} T_f(r) &\geq \int_{t_0}^{r} \frac{dt}{t} \int_{\Delta_{s,t_0}} \Omega_0 + \int_{t_0}^{r} \frac{dt}{t} \int_{\Delta_{t_0,r}} \Omega_0 \\ &\geq \int_{t_0}^{r} \frac{dt}{t} \int_{\Delta_{s,t_0}} \Omega_0 \\ &\geq K \log \frac{r}{t_0}, \end{aligned}$$

which tends to $+\infty$ as r tends to $+\infty$.

PROPOSITION 3.3.2. *$T_f(r)$ is an increasing function on $[s, +\infty)$ which is convex with respect to $\log r$.*

PROOF. As in the proof of Proposition 3.2.4, if we set $\Omega_k = h_k dd^c |z|^2$, we have

$$\frac{dT_f(r)}{d\log r} = r\frac{dT_f(r)}{dr} = \frac{1}{\pi}\int_s^r \rho d\rho \int_0^{2\pi} h_k(\rho e^{i\theta}) d\theta \geq 0,$$

$$\frac{1}{r^2}\frac{d^2 T_f(r)}{(d\log r)^2} = \frac{1}{\pi}\int_0^{2\pi} h_k(re^{i\theta}) d\theta \geq 0.$$

This gives Proposition 3.3.2.

By a coordinate change $z = 1/u, \Delta_{s,\infty}$ is transformed to the set $\bar{\Delta}^*_{1/s} := \{z; 0 < |z| \leq 1/s\}$. We say that f has an *essential singularity* at ∞ if the map $g(u) = f(1/u) : \Delta^*_{1/s} \to P^n(\mathbf{C})$ cannot be holomorphically continued to the disc $\Delta_{1/s}$.

PROPOSITION 3.3.3. *For a holomorphic curve f in $P^n(\mathbf{C})$ defined on a neighborhood of $\Delta_{s,\infty}$, the following four conditions are equivalent.*

(i) *f has an essential singularity at ∞.*

(ii) *There is a hyperplane H in $P^n(\mathbf{C})$ such that $f^{-1}(H)$ contains infinitely many points.*

(iii) *For some hyperplane H in $P^n(\mathbf{C})$, $\lim_{r\to\infty} \dfrac{N_f(r,H)}{\log r} = +\infty$.*

(iv) $\lim_{r\to\infty} \dfrac{T_f(r)}{\log r} = +\infty$.

PROOF. Assume that (i) is true. Let $f = (f_0 : \cdots : f_n)$ be a reduced representation of f. After a suitable change of indices, if necessary, we may assume that $f_0 \not\equiv 0$ and $f_1/f_0 (= (f_0 : f_1)) : \Delta_{s,\infty} \to P^1(\mathbf{C})$ is a meromorphic function with an essential singularity at ∞. For each integer $\ell > s$ we set $F_\ell := P^1(\mathbf{C}) - f(\Delta_{\ell,\infty})$, which is a nowhere dense closed set as a result of the classical Casorati-Weierstrass theorem. By the category theorem we can choose a value $a = (a_0 : a_1)$ in $P^1(\mathbf{C})$ which is not contained in any F_ℓ. Then, for a hyperplane

$$H : a_1 w_0 - a_0 w_1 = 0,$$

we have $f^{-1}(H) \cap \Delta_{\ell,\infty} \neq \emptyset$ for all ℓ and so $f^{-1}(H)$ contains infinitely many points. We then have (ii).

We next assume that (ii) is valid. Then, for any large number k there is some r_0 such that

$$n(r, \nu(f,H)) \geq k \qquad \text{for all } r > r_0.$$

Therefore, we see

$$N_f(r,H) = \int_s^r \frac{n(t,\nu(f,H))}{t} dt \geq k \log \frac{r}{r_0},$$

and so

$$\liminf_{r\to\infty} \frac{N_f(r,H)}{\log r} \geq k.$$

Since k is arbitrary, we have (iii).

The condition (iv) is a direct result of (iii) in view of Corollary 3.1.16.

Lastly, assume that f has a removable singularity at ∞. Then the map $g(u) := f(1/u)$ has a reduced representation $g = (g_0 : \cdots : g_n)$ in a neighborhood of $u = 0$. The original map f has a reduced representation $f = (h_0 : \cdots : h_n)$ for the functions $h_i(z) = g_i(1/z)$. Consider the function

$$|F| := (|h_0|^2 + \cdots + |h_n|^2)^{1/2}.$$

Since the h_i are bounded near ∞, from Corollary 3.1.12 we can conclude that

$$T_f(r) \leq \frac{1}{2\pi} \int_0^{2\pi} \log |F(re^{i\theta})| d\theta + O(\log r) \leq O(\log r).$$

This completes the proof of Proposition 3.3.3.

COROLLARY 3.3.4. *Let f be a nondegenerate holomorphic map of $\mathbf{C}$ into $P^n(\mathbf{C})$. The map f is rational, namely, f is representable as $f = (f_0 : f_1 : \cdots : f_n)$ with polynomials f_i if and only if*

$$\lim_{r\to\infty} \frac{T_f(r)}{\log r} < +\infty.$$

PROOF. This is due to the fact that a meromorphic function φ on $\mathbf{C}$ is rational if and only if φ has a removable singularity at ∞.

DEFINITION 3.3.5. Let H be a hyperplane in $P^n(\mathbf{C})$ with $f(\Delta_{s,\infty}) \not\subseteq H$ and m a positive integer or $+\infty$. We define the *defect* (truncated by m) of H for f by

$$\delta_f(H)^{[m]} = 1 - \limsup_{r\to\infty} \frac{N_f(r,H)^{[m]}}{T_f(r)}.$$

For convenience's sake, we set $\delta_f(H)^{[m]} = 0$ if $f(\Delta_{s,\infty}) \subseteq H$ and, for brevity, we denote $\delta_f(H)^{[\infty]}$ by $\delta_f(H)$.

PROPOSITION 3.3.6. *If $s = 0$ or if f has an essential singularity at ∞, then*

$$0 \leq \delta_f(H)^{[m]} \leq 1$$

for every hyperplane H in $P^n(\mathbf{C})$.

PROOF. By Corollary 3.1.16, we have

$$\frac{N_f(r,H)^{[n]}}{T_f(r)} + \frac{m_f(r,H)}{T_f(r)} \leq 1 + \frac{\eta_s(r)}{T_f(r)},$$

where

$$\lim_{r\to\infty} \frac{\eta_s(r)}{T_f(r)} = 0$$

by Propositions 3.3.1 or 3.3.3. This gives Proposition 3.3.6.

The following proposition gives some geometric meanings of the defect.

PROPOSITION 3.3.7. *Let $f : \Delta_{s,\infty} \to P^n(\mathbf{C})$ be a nondegenerate holomorphic map and H a hyperplane in $P^n(\mathbf{C})$.*

(i) *If $f^{-1}(H) = \phi$, then*

$$\delta_f(H) = 1.$$

(ii) *If f has an essential singularity at ∞ and $\# f^{-1}(H) < +\infty$, then*

$$\delta_f(H) = 1.$$

(iii) *Assume that $s = 0$ or that f has an essential singularity at ∞. For some $m(> n)$, if $\nu(f, H)(\zeta) \geq m$ for every ζ in $f^{-1}(H)$, then*

$$\delta_f(H)^{[n]} \geq 1 - \frac{n}{m}.$$

PROOF. The assertion (i) is obvious because the counting function vanishes identically. If f has an essential singularity at ∞ and $f^{-1}(H)$ is finite, then we see easily that $N_f(r, H) = O(\log r)$ and so

$$\limsup_{r\to\infty} \frac{N_f(r,H)}{T_f(r)} = \limsup_{r\to\infty} \frac{N_f(r,H)}{\log r} \frac{\log r}{T_f(r)} = 0.$$

This gives (ii).

Assume that $\nu(f, H)(\zeta) \geq m$ for every $\zeta \in f^{-1}(H)$. Then we have

$$\min(\nu_f(r,H), n) \leq n \min(\nu_f(r,H), 1) \leq \frac{n}{m} \nu_f(r,H)$$

and so

$$N_f(r,H)^{[n]} \leq n N_f(r,H)^{[1]} \leq \frac{n}{m} N_f(r,H).$$

Therefore,

$$1 - \frac{N_f(r,H)^{[n]}}{T_f(r)} \geq 1 - \frac{n}{m} \frac{N_f(r,H)}{T_f(r)},$$

whence the desired inequality follows from Corollary 3.1.16.

We now give the following defect relation for a holomorphic curve.

THEOREM 3.3.8. *Let $f : \Delta_{s,\infty} \to P^n(\mathbf{C})$ be a nondegenerate holomorphic map. If $s = 0$ or else if f has an essential singularity at ∞, then for* *arbitrarily given hyperplanes $H_1, \dots, H_q$ located in N-subgeneral position we have*

$$\sum_{j=1}^{q} \omega(j)\delta_f(H_j)^{[n]} \leq n+1,$$

where $\omega(j)$ are Nochka weights for H_j's.

PROOF. For every $\varepsilon > 0$, using Theorems 3.2.12 and 2.4.11, (ii), we have

$$\sum_{j=1}^{q} \omega(j)(T_f(r) - N_f(r,H_j)^{[n]}) \leq (n+1)T_f(r) + \theta\varepsilon T_f(r) + O(\log T_f(r)) + \eta_s(r)$$

and so

$$\sum_{j=1}^{q} \omega(j)\left(1 - \frac{N_f(r,H_j)^{[n]}}{T_f(r)}\right) \leq n+1+\frac{O(\log T_f(r))}{T_f(r)} + \frac{\eta_s(r)}{T_f(r)} + \theta\varepsilon$$

for all r except on a set E with finite logarithmic measure. Taking lower limits of both sides of this inequality, we easily conclude Theorem 3.3.8 by the use of Propositions 3.3.1 or 3.3.3 because ε may be chosen arbitrarily small.

COROLLARY 3.3.9. *Let $f : \Delta_{s,\infty} \to P^n(\mathbf{C})$ be a nondegenerate holomorphic map. If $s = 0$ or else if f has an essential singularity at ∞, then for arbitrarily given hyperplanes $H_1, \dots, H_q$ located in N-subgeneral position we have*

$$\sum_{j=1}^{q} \delta_f(H_j)^{[n]} \leq 2N - n + 1.$$

PROOF. By the use of Theorem 2.4.11, (i) and (ii), Theorem 3.3.8 implies that

$$\begin{aligned}\sum_{j=1}^{q} \delta_f(H_j)^{[n]} &= q + \sum_{j=1}^{q}(\delta_f(H_j)^{[n]} - 1)\\ &\leq q + \frac{1}{\theta}\sum_{j=1}^{q} \omega(j)(\delta_f(H_j)^{[n]} - 1)\\ &\leq q + \frac{1}{\theta}\left(n+1-\sum_{j=1}^{q}\omega(j)\right)\\ &= 2N - n + 1.\end{aligned}$$

This gives Corollary 3.3.9.

We give here some applications of the defect relation.

THEOREM 3.3.10. *For a holomorphic map $f : \mathbf{C} \to P^n(\mathbf{C})$, if f omits $2N - n + 2$ hyperplanes in $P^n(\mathbf{C})$ located in N-subgeneral position, then f is degenerate.*

PROOF. Assume that f is nondegenerate and omits q hyperplanes $H_j (1 \leq j \leq q)$ located in N-subgeneral position. Then, by Proposition 3.3.7 and Corollary 3.3.9, we obtain

$$q = \sum \delta_f(H_j)^{[n]} \leq 2N - n + 1.$$

This contradicts the assumption. Therefore f is necessarily degenerate.

COROLLARY 3.3.11. *Let f be a holomorphic map of $\mathbf{C}$ into $P^N(\mathbf{C})$ such that the image of f is not included in any projective linear subspace of dimension less than n. Then f can omit at most $2N - n + 1$ hyperplanes in $P^N(\mathbf{C})$ located in general position.*

PROOF. Consider the smallest projective linear subspace of $P^N(\mathbf{C})$ containing $f(\mathbf{C})$, which we denote by $P^m(\mathbf{C})$ if it is of dimension m. By assumption, $m \geq n$ and f is considered as a nondegenerate holomorphic map into $P^m(\mathbf{C})$. The intersection of $P^m(\mathbf{C})$ with the given hyperplanes located in general position in $P^N(\mathbf{C})$ are in N-subgeneral position in $P^m(\mathbf{C})$. Corollary 3.3.11 is then an immediate consequence of Theorem 3.3.10.

COROLLARY 3.3.12. *For a holomorphic map $f : \mathbf{C} \to P^n(\mathbf{C})$, if f omits $n + 2$ hyperplanes in $P^n(\mathbf{C})$ located in general position, then f is degenerate.*

PROOF. Assume that f is nondegenerate. Since general position means n-subgeneral position, f can omit at most $n + 1$ hyperplanes in general position by Theorem 3.3.10. Corollary 3.3.12 is a restatement of this fact.

We have also the following small Picard theorem which is a generalization of Theorem 1.5.3.

COROLLARY 3.3.13. *For a holomorphic map $f : \mathbf{C} \to P^N(\mathbf{C})$, if f omits $2N + 1$ hyperplanes in $P^N(\mathbf{C})$ located in general position, then f is a constant.*

PROOF. Assume that f is not a constant. Then, for $n = 1$, f satisfies the assumption of Corollary 3.3.11. Therefore, f cannot omit $2N + 1$ hyperplanes in general position. This gives Corollary 3.3.13.

Using Proposition 3.3.7, (ii), we can also prove the following generalization of Theorem 1.5.12 by the same argument as in Theorem 3.3.10.

THEOREM 3.3.14. *Let $f : \Delta_{s,\infty} \to P^n(\mathbf{C})$ be a nondegenerate holomorphic map which has an essential singularity at ∞. If there are $n+2$ hyperplanes $H_1, \dots, H_{n+2}$ in $P^n(\mathbf{C})$ located in general position such that the inverse images $f^{-1}(H_j)(1 \leq j \leq n+2)$ are finite, then f is degenerate.*

THEOREM 3.3.15. *Let $f : \Delta_{s,\infty} \to P^n(\mathbf{C})$ be a nondegenerate holomorphic map and $H_1, \dots, H_q$ hyperplanes in N-subgeneral position. Assume that f has an essential singularity at ∞ in the particular case $s > 0$, and that there are positive integers $m_1, \dots, m_q$ such that $m_j > n$ and $\nu(f, H_j)(\zeta) \geq m_j$ for every point $\zeta \in f^{-1}(H_j)(1 \leq j \leq q)$. Then,*

$$\sum_{j=1}^{q} \left(1 - \frac{n}{m_j}\right) \leq 2N - n + 1.$$

PROOF. This is an immediate consequence of Corollary 3.3.9 and Proposition 3.3.7, (iii).

§3.4 Borel's theorem and its applications

In this section, we give some applications of the defect relation and, in the next two sections we prove the defect relation for the derived curves of a holomorphic curve in $P^n(\mathbf{C})$. These are not used in the following chapters. Readers who are interested only in minimal surfaces may skip these three sections.

By $\mathcal{M}$ we denote the field of all meromorphic functions on $\Delta_{s,\infty}(:= \{z; s \leq |z| < +\infty\})$ which have no essential singularities at ∞. We give the following improvement of the classical Borel's theorem(cf. [20]).

THEOREM 3.4.1. *Let $f_1, \dots, f_p$ $(p \geq 2)$ be nonzero holomorphic functions on $\Delta_{s,\infty}$. Suppose that there exist positive integers $m_1, \dots, m_p$ such that*

(i) $\nu_{f_i}(\zeta) \geq m_i$ $(1 \leq i \leq p)$ *whenever ζ is a zero of f_i,*

(ii) *if $f_{i_0}, \dots, f_{i_k}(1 \leq i_0 < i_1 < \cdots < i_k \leq p)$ have a common zero of multiplicities $n_{i_0}, \dots, n_{i_k}$ respectively, then*

$$n'_{n_\ell} := n_{i_\ell} - \min(n_{i_0}, n_{i_1}, \dots, n_{i_k}) \geq m_{i_\ell}$$

for any ℓ *with* $n'_{i_\ell} > 0$ *and*

(iii) $\sum_{i=1}^{p} \frac{p-2}{m_i} < 1$

(iv) f_i/f_j *have essential singularities at* ∞ *for* $i \neq j$.

Then, $f_1, f_2, \dots, f_p$ *are linearly independent over the field* $\mathcal{M}$.

PROOF. The proof is given by induction on p. For the case $p = 2$ the conclusion is obviously true under only the assumption (iv). Assume that Theorem 3.4.1 is true for the cases $\leq p-1$, where $p \geq 3$. It suffices to show that an arbitrarily given linear relation

$$\alpha_1 f_1 + \cdots + \alpha_p f_p = 0 \qquad (\alpha_i \in \mathcal{M}) \tag{3.4.2}$$

yields that at least one α_i vanishes.

Suppose that $\alpha_i \not\equiv 0$ for all i. Choose a nonzero meromorphic function g on $\Delta_{s,\infty}$ such that

$$\nu_g = \min(\nu_{f_1}, \dots, \nu_{f_{p-1}}).$$

Set $g_i := f_i/g$ $(1 \leq i \leq p-1)$. These are holomorphic functions without common zeros and satisfy the conditions (i) $\sim$ (iv) of Theorem 3.4.1. Therefore, they are linearly independent over $\mathcal{M}$ by the induction hypothesis.

We now set $h_i := \alpha_{i+1} g_{i+1} (0 \leq i \leq p-2), h_{p-1} := -\alpha_p f_p/g$ and define a holomorphic map

$$h = (h_0 : \cdots : h_{p-2}) : \Delta_{s,\infty} \to P^{p-2}(\mathbf{C}).$$

Then, as is stated above, h is nondegenerate. Consider p hyperplanes

$$\begin{aligned} H_j \quad &: w_j = 0 \qquad\qquad (0 \leq j \leq p-2) \\ H_{p-1} &: w_0 + \cdots + w_{p-2} = 0, \end{aligned}$$

which are located in general position. By the assumption (iv), h has an essential singularity at ∞. On the other hand, since α_i does not have essential singularities at ∞, we can take a number s_0 such that α_i has neither a zero nor a pole in $\Delta_{s_0,\infty}$. As a result of Proposition 3.3.7, (iii) and the assumption (i), we have

$$\delta_h(H_j) \geq 1 - \frac{p-2}{m_j}.$$

On the other hand, by Theorem 3.3.8 we get

$$\sum_{j=1}^{p}\left(1-\frac{p-2}{m_j}\right)\leq p-1,$$

which contradicts the assumption (iv). This completes the proof of Theorem 3.4.1.

Now, we give some applications of Theorem 3.4.1.

PROPOSITION 3.4.3. *Let $f_1, f_2, \ldots, f_p$ be holomorphic functions on $\Delta_{s,\infty}$ such that*
(i) *each f_i is nowhere vanishing,*
(ii) *f_i/f_j has an essential singularity at ∞ for $i \neq j$.*
Then $f_1, \ldots, f_p$ are linearly independent over $\mathcal{M}$.

PROOF. Take sufficiently large m_i satisfying the condition (iii) of Theorem 3.4.1. Since f_i has no zero, the conditions (i) and (ii) of Theorem 3.4.1 are also satisfied. Thus we have Proposition 3.4.3.

COROLLARY 3.4.4. *Let $f_1, \ldots, f_p$ be nowhere vanishing entire functions such that f_i/f_j are not constants for any distinct i and j. Then they are linearly independent over* $\mathbf{C}$.

PROOF. As is easily seen, a nowhere vanishing entire function is necessarily a constant if it dose not have an essential singularity at ∞. Corollary 3.4.4 is now an immediate consequence of Proposition 3.4.3.

PROPOSITION 3.4.5. *Let $\eta_1, \ldots, \eta_p$ be nowhere zero holomorphic functions on $\Delta_{s,\infty}$. Assume that*

$$\eta_1^{\ell_1}\eta_2^{\ell_2}\ldots\eta_p^{\ell_p}\notin\mathcal{M}$$

for any integers $\ell_1, \ldots, \ell_p$ with $(\ell_1, \ldots, \ell_p) \neq (0, \ldots, 0)$. Then, for any nonzero polynomial $P(X_1, \ldots, X_p)$ with coefficients in $\mathcal{M}$, we have

$$P(\eta_1, \eta_2, \ldots, \eta_p) \not\equiv 0.$$

PROOF. Assume that some polynomial

$$P(X_1, X_2, \ldots, X_p) = \sum_{\ell_1,\ldots,\ell_p} \alpha_{\ell_1\ldots\ell_p} X_1^{\ell_1}\ldots X_p^{\ell_p} \qquad (\alpha_{\ell_1\ldots\ell_p}\in\mathcal{M})$$

satisfies the condition

$$\sum_{\ell_1,\dots,\ell_p} \alpha_{\ell_1\dots\ell_p}\eta_1^{\ell_1}\dots\eta_p^{\ell_p} \equiv 0.$$

By assumption,

$$\frac{\eta_1^{\ell_1}\dots\eta_p^{\ell_p}}{\eta_1^{m_1}\dots\eta_p^{m_p}} \notin \mathcal{M}$$

for any distinct $(\ell_1,\dots,\ell_p)$ and $(m_1,\dots,m_p)$. According to Proposition 3.4.3, $\{\eta_1^{\ell_1}\cdots\eta_p^{\ell_p}\}$ are linearly independent over $\mathcal{M}$. This implies $P(X_1,\dots,X_p)\equiv 0$ and so Proposition 3.4.5.

Since a nowhere zero holomorphic function in $\mathcal{M}$ is necessarily a constant for the case $s=0$, we have easily the following:

COROLLARY 3.4.6. *Let $f_1,\dots,f_p$ be nowhere zero holomorphic functions on $\mathbf{C}$ which are multiplicatively independent over $\mathbf{Z}$ up to a nonzero constant multiple. Then they are algebraically independent.*

PROPOSITION 3.4.7. *Let $f_1,\dots,f_p$ be nonzero meromorphic functions on $\Delta_{s,\infty}$ which satisfy the conditions* (i) $\sim$ (iii) *of Theorem* 3.4.1 *for some positive integers $m_i(1\leq i\leq p)$. Assume that*

$$\alpha_1 f_1+\cdots+\alpha_p f_p=0$$

for some nonzero functions $\alpha_i\in\mathcal{M}$. Consider the partition

$$\{1,2,\dots,p\}=I_1\cup I_2\cup\ldots\cup I_k$$

such that i and j are in the same class I_ℓ if and only if $f_i/f_j\in\mathcal{M}$. Then

$$\sum_{i\in I_\ell}\alpha_i f_i\equiv 0$$

for any ℓ.

PROOF. For an arbitrarily chosen $i_\ell\in I_\ell$, since $f_i/f_{i_\ell}\in\mathcal{M}$ for any $i\in I_\ell$, we can write

$$\sum_{i=1}^{p}\alpha_i f_i=\sum_{\ell=1}^{k}\sum_{i\in I_\ell}\alpha_i f_i=\sum_{\ell=1}^{k}\gamma_\ell f_{i_\ell}=0$$

with some $\gamma_\ell \in \mathcal{M}$. By the definition of the partition, f_{i_ℓ}/f_{i_m} have essential singularities at ∞ for any mutually distinct ℓ and m. We apply Theorem 3.4.1 to the functions $f_{i_1}, \ldots, f_{i_k}$. We can conclude that they are linearly independent over $\mathcal{M}$ and so $\gamma_\ell = 0$ for all ℓ. This implies that

$$\sum_{i \in I_\ell} \alpha_i f_i = \gamma_\ell f_{i_\ell} = 0.$$

The proof of Proposition 3.4.7 is completed.

For a positive integer d the hypersurface

$$V_d := \{(w_0 : \cdots : w_{n+1}); w_0^d + \cdots + w_{n+1}^d = 0\}$$

in $P^{n+1}(\mathbf{C})$ is called the *Fermat variety* of degree d.

COROLLARY 3.4.8. *Let $d > n(n+2)$. For a holomorphic map f of $\mathbf{C}$ into $P^{n+1}(\mathbf{C})$, if $f(\mathbf{C}) \subseteq V_d$, then the image is necessarily included in an at most $[n/2]$-dimensional subvariety of V_d, where $[a]$ denotes the largest integer which is not larger than a.*

PROOF. Take a reduced representation $f = (f_0 : \cdots : f_{n+1})$. We may assume $f_i \not\equiv 0$ for any i. By assumption,

$$f_0^d + \cdots + f_{n+1}^d = 0.$$

Set $g_i := f_{i-1}^d$ for $i = 1, \ldots, n+2$. The values of the divisors ν_{g_i} $(0 \leq i \leq n)$ are multiples of d everywhere. Set $p = n+2$ and $m_i = d$ for each i $(0 \leq i \leq n+1)$. Since

$$\sum_{i=1}^{p} \frac{1}{m_i} = \frac{n+2}{d} < \frac{1}{n} = \frac{1}{p-2},$$

the functions $g_1, \ldots, g_p$ satisfy the assumptions of Proposition 3.4.7 for $\alpha_i \equiv 1$. Changing indices if necessary, we may assume that, for the sets

$$I_1 := \{1, \ldots, p_1\}, I_2 := \{p_1 + 1, \ldots, p_2\}, \ldots, I_k := \{p_{k-1} + 1, \ldots, p_k\},$$

i and j are in the same class if and only if $g_i/g_j \in \mathcal{M}$ and

$$\sum_{i \in I_\ell} g_i = 0.$$

Consider the subvariety

$$W := \{(w_0 : \cdots : w_{n+1}) \in V_d; \sum_{i \in I_\ell} w_{i-1}^d = 0 \text{ for all } \ell = 1, 2, \ldots, k\},$$

which includes the image of f. Since g_i is not zero, the number of elements of each I_ℓ is larger than one and so $k \leq p/2 = (n+2)/2$. Therefore, the dimension of W is not larger than $n/2$. This completes the proof of Corollary 3.4.8.

In 1926, R. Nevanlinna showed that, for two nonconstant meromorphic functions φ and ψ on $\mathbf{C}$, $\varphi^{-1}(\alpha_j) = \psi^{-1}(\alpha_j)$ for five distinct values $\alpha_1, \ldots, \alpha_5$ only if $\varphi \equiv \psi$([54]). Related to this result, we can prove the following uniqueness theorem([21]).

THEOREM 3.4.9. *Let f, g be nonconstant holomorphic maps of $\mathbf{C}$ into $P^n(\mathbf{C})$ at least one of which is nondegenerate. If there exist $3n+2$ hyperplanes $H_1, \ldots, H_{3n+2}$ located in general position such that $f(\mathbf{C}) \not\subset H_j$, $g(\mathbf{C}) \not\subset H_j$ and $\nu(f, H_j) = \nu(g, H_j)$ for $j = 1, \ldots, 3n+2$, then $f \equiv g$.*

PROOF. Set $q := 3n + 2$. We write given hyperplanes H_j as

$$H_j : a_{j0}w_0 + \cdots + a_{jn}w_n = 0 \qquad (j = 1, \ldots, q).$$

Taking reduced representations $f = (f_0 : \cdots : f_n)$ and $g = (g_0 : \cdots : g_n)$, we set

$$F_j^f := a_{j0}f_0 + \cdots + a_{jn}f_n, \quad F_j^g := a_{j0}g_0 + \cdots + a_{jn}g_n$$

and define functions

$$h_j := F_j^g / F_j^f \qquad (j = 1, 2, \ldots, q), \tag{3.4.10}$$

which are nowhere zero holomorphic functions on $\mathbf{C}$. We choose an arbitrary subset Q of the index set $Q_0 := \{1, 2, \ldots, q\}$ with $\#Q = 2n + 2$.

We now prove that

(3.4.11) *For each $I \subset Q$ there exists some $J \subset Q$ with $I \neq J$ such that h_I / h_J is a constant, where $h_I := h_{i_0} h_{i_1} \cdots h_{i_n}$ for each subset $I := \{i_0, \ldots, i_n\}$ of Q.*

To this end, there is no loss of generality in assuming that $Q = \{1, 2, \ldots, 2n+2\}$. We rewrite (3.4.10) as

$$a_{j0}g_0 + \cdots + a_{jn}g_n = h_j(a_{j0}f_0 + \cdots + a_{jn}f_n) \qquad (j \in Q).$$

Eliminating $2n+2$ functions $f_0, \dots, f_n$ and $g_0, \dots, g_n$ from these $2n+2$ identities, we get

$$\Psi := \det(a_{j0}, \dots, a_{jn}, a_{j0}h_j, \dots, a_{jn}h_j; j = 1, 2, \dots, 2n+2) = 0.$$

For each system $I := \{i_0, \dots, i_n\}$ with $1 \le i_0 < \dots < i_n \le q$, we set

$$\varepsilon_I := (-1)^{\frac{n(n+1)}{2} + i_0 + \dots + i_n},$$

$$A_I := \det\left(a_{i_r j}; {}^{0 \le r \le n}_{0 \le j \le n}\right) \det\left(a_{i'_s j}; {}^{0 \le s \le n}_{0 \le j \le n}\right),$$

where $i'_0, \dots, i'_n$ are the indices with $\{i_0, \dots, i_n, i'_0, \dots, i'_n\} = Q$ and $1 \le i'_0 < \dots < i'_n \le 2n+2$. Then, by the Laplace expansion theorem, we have

$$\Psi = \sum_{I \subset Q, \# I = n+1} \varepsilon_I A_I h_I \tag{3.4.12}$$

Since all $n+1$ vectors among $\{(a_{j0}, \dots, a_{jn}); 1 \le j \le 2n+2\}$ are linearly independent, we have $A_I \ne 0$ for all I's. We can apply Proposition 3.4.7 to the identity (3.4.12) to obtain the desired conclusion (3.4.11).

We next consider the multiplicative group $\mathcal{H}^*$ of all nowhere zero holomorphic functions on $\mathbf{C}$, which includes the subgroup $\mathbf{C}^* := \mathbf{C} - \{0\}$ consisting of all nonzero constant functions. Then, the factor group $\mathcal{G} := \mathcal{H}^*/\mathbf{C}^*$ is a torsion free abelian group. We denote by $[h]$ the class in $\mathcal{G}$ containing $h \in \mathcal{H}^*$. Consider the subgroup $\tilde{\mathcal{G}}$ of $\mathcal{G}$ generated by $[h_1], \dots, [h_q]$ and choose suitable functions $\eta_1, \dots, \eta_t \in \mathcal{H}^*$ such that $[\eta_1], \dots, [\eta_t]$ give a basis of $\tilde{\mathcal{G}}$. Then, each h_j can be uniquely represented as

$$h_j = c_j \eta_1^{\ell_{j1}} \eta_2^{\ell_{j2}} \cdots \eta_t^{\ell_{jt}}$$

with some nonzero constant c_j and integers $\ell_{j\tau}$.

(3.4.13) *For these integers $\ell_{j\tau}$ we can choose suitable integers $p_1, \dots, p_t$ satisfying the condition that, for integers*

$$\ell_j := \ell_{j1}p_1 + \dots + \ell_{jt}p_t \qquad (1 \le j \le q),$$

$\ell_i = \ell_j$ if and only if $(\ell_{i1}, \dots, \ell_{it}) = (\ell_{j1}, \dots, \ell_{jt})$, or equivalently h_i/h_j is a constant.

In fact, this is trivial for the case $t = 1$. Moreover, if we can choose $p_1, \dots, p_{t-1}$ satisfying the same condition for the vectors $(\ell_{j1}, \dots, \ell_{jt-1})$, it is easy to show that there are only finitely many integers p_t such that $p_1, \dots, p_t$ do not satisfy the desired condition.

We shall show that:

(3.4.14) *there is a subset I_0 of Q_0 with $\#I_0 = n+2$ such that h_i/h_j are constants for all $i, j \in I_0$.*

To this end, we assume that, after a suitable change of indices,

$$\ell_1 \le \ell_2 \le \cdots \le \ell_q$$

for the above integers ℓ_j. By virtue of (3.4.13) we have only to show that

$$\ell_1 \le \cdots \le \ell_{n+1} = \cdots = \ell_{2n+2} \le \cdots \le \ell_{3n+2}. \tag{3.4.15}$$

To see this, take the subset $Q := \{1, \dots, n+1, 2n+2, \dots, 3n+2\}$ of Q_0 which contains $2n+2$ elements and apply (3.4.11) to the h_j's ($j \in Q$) to show that there is a subset $\{i_0, \dots, i_n\}$ of Q satisfying the condition that $\{i_0, \dots, i_n\} \neq \{1, \dots, n+1\}$ and

$$\frac{h_{i_0} h_{i_1} \cdots h_{i_n}}{h_1 h_2 \cdots h_{n+1}} \in \mathbf{C}.$$

By (3.4.13), we then have

$$\begin{aligned}(\ell_{i_0} - \ell_1) + \cdots + (\ell_{i_n} - \ell_{n+1})\\ = \ell_{i_0} + \ell_{i_1} + \cdots + \ell_{i_n} - (\ell_1 + \cdots + \ell_{n+1}) = 0.\end{aligned}$$

Since $\ell_{i_0} \ge \ell_1, \dots, \ell_{i_n} \ge \ell_{n+1}$ and $i_n \ge 2n+2$, this is possible only when $\ell_{n+1} = \ell_{i_n} (= \ell_{2n+2})$. This concludes (3.4.14).

As a consequence of (3.4.14), we may assume that h_i/h_1 are all constants for $i = 2, 3, \dots, n+2$ after a suitable change of indices. Moreover, it may be assumed that f is nondegenerate and the given hyperplanes $H_j (1 \le j \le n+2)$ are represented as

$$\begin{aligned} H_j \quad &: w_{j-1} = 0 \qquad\qquad (1 \le j \le n+1)\\ H_{n+2} &: w_0 + \cdots + w_n = 0\end{aligned}$$

after a suitable change of homogeneous coordinates. Then, we may rewrite (3.4.10) as $g_j = c_j h_1 f_j (0 \le j \le n)$ and

$$g_0 + \cdots + g_n = c_{n+1} h_1 (f_0 + \cdots + f_n)$$

for some $c_{n+1} \neq 0$. These imply that

$$(c_{n+1} - c_0) f_0 + \cdots + (c_{n+1} - c_n) f_n = 0.$$

By the nondegeneracy of f, we conclude that $c_0 = \cdots = c_n = c_{n+1}$. This gives the desired conclusion $f \equiv g$.

We can also prove the following theorem.

THEOREM 3.4.16. *Let f, g be nonconstant holomorphic maps of $\mathbf{C}$ into $P^n(\mathbf{C})$. Assume that there exist $2n+3$ hyperplanes $H_1, \dots, H_{2n+3}$ located in general position such that $f(\mathbf{C}) \not\subset H_j$, $g(\mathbf{C}) \not\subset H_j$ and $\nu(f, H_j) = \nu(g, H_j)$ for $j = 1, \dots, 2n+3$. If f or g is algebraically nondegenerate, namely the image is not contained in any proper algebraic subset of $P^n(\mathbf{C})$, then $f \equiv g$.*

For the proof, see [22] and [23].

For other results relating to uniqueness problems for holomorphic curves, refer to [24], [25], [26] and [33]. Moreover, we can prove some finiteness theorems for holomorphic curves. For these topics, refer to [31] and [34].

§3.5 Some properties of Wronskians

In the next section, we shall prove defect relations for the derived curves of a holomorphic curve in $P^n(\mathbf{C})$. For this purpose, we give some preliminary properties of Wronskians associated with the derived curves of a holomorphic curve in $P^n(\mathbf{C})$.

Let $f : \mathbf{C} \to P^n(\mathbf{C})$ be a nondegenerate holomorphic curve and let $f^k : \mathbf{C} \to P^{N_k}(\mathbf{C})$ $(0 \le k \le n-1)$ be the derived curves of f. We first give the following:

DEFINITION 3.5.1. We say a hyperplane H is *decomposable* if H is represented as

$$H : \langle W, A^k \rangle = 0$$

with a nonzero decomposable k-vector $A^k \in \bigwedge^{k+1} \mathbf{C}^{n+1}$.

The derived curves f^k are not necessarily nondegenerate even if f is nondegenerate. Here, we note the following fact.

(3.5.2). *If a holomorphic curve $f : \mathbf{C} \to P^n(\mathbf{C})$ is nondegenerate, then the images $f^k(\mathbf{C})$ are not included in any decomposable hyperplane in $P^{N_k}(\mathbf{C})$.*

PROOF. For a given decomposable unit vector A^k we can choose a system of orthonormal basis $E_0, \dots, E_n$ of $\mathbf{C}$ such that $A^k = E_0 \wedge \dots \wedge E_k$. Using these coordinate system, we represents f as $f = (f_0 : \dots : f_n)$. Then, for the map

$$F_k := F \wedge F' \wedge \dots \wedge F^{(k)} : \mathbf{C} \to \mathbf{C}^{N_k+1}$$

we have

$$\langle F_k, A^k \rangle \equiv W(f_0, \dots, f_k) \not\equiv 0$$

as a result of Proposition 2.1.7 because $f_0, \dots, f_k$ are linearly independent, where $W(f_0, \dots, f_k)$ denotes the Wronskian of $f_0, \dots, f_k$. This shows (3.5.2).

For given nonnegative integers r, k, we use the following notation:

$$\mathcal{I}_r^k := \{(i_0, \dots, i_r); 0 \leq i_0 < \dots < i_r \leq k\},$$
$$\mathcal{I}_r := \{(i_0, \dots, i_r); 0 \leq i_0 < \dots < i_r < \infty\}.$$

We denote a particular element $(0, 1, \dots, r) \in \mathcal{I}_r$ by $I_{0,r}$, or simply I_0 if we have no reason for confusion.

We take a reduced representation $f = (f_0 : \dots : f_n)$ on $\mathbf{C}$. We shall study holomorphic functions $f_0, \dots, f_n$. For $I := (i_0, \dots, i_r) \in \mathcal{I}_r$ and $J := (j_0, \dots, j_r) \in \mathcal{I}_r^k$, we set

$$W(I, J) \equiv W(i_0, \dots, i_r; j_0, \dots, j_r) := \det\left(f_{j_m}^{(i_\ell)}; 0 \leq \ell, m \leq r\right).$$

In particular, $W(0, 1, \dots, r; j_0, \dots, j_r)$ means the Wronskian of the functions $f_{j_0}, \dots, f_{j_r}$.

DEFINITION 3.5.3. For each $I = (i_0, \dots, i_r) \in \mathcal{I}_r$ we define the *weight* of I by

$$w(I) := (i_0 - 0) + (i_1 - 1) + \dots + (i_r - r).$$

DEFINITION 3.5.4. Consider a polynomial $P(u_1, u_2, \dots, u_r)$ in $u_1, u_2, \dots, u_r$ for which a weight w_ℓ is associated with each variable u_ℓ. We call P *isobaric* of weight w if $P(t_1^{w_1}, t_2^{w_2}, \dots, t_r^{w_r})$ is a homogeneous polynomial of degree w with respect to the variables $t_1, \dots, t_r$.

We now give the following:

LEMMA 3.5.5. *For every $I \in \mathcal{I}_k$, the meromorphic function $\dfrac{W(I; I_{0,k})}{W(I_{0,k}; I_{0,k})}$ can be written as a polynomial of several functions of the type*

$$\left(\frac{W(I_{0,r}; J)'}{W(I_{0,r}; J)}\right)^{(\ell-1)}, \tag{3.5.6}$$

where $0 \leq r \leq k$, $\ell = 1, 2, \dots$ and $J \in \mathcal{I}_r^k$.

If we associate the weight ℓ with the function given by (3.5.6), such a polynomial can be chosen so as to be isobaric of weight $w(I)$.

PROOF. We shall give the proof of Lemma 3.5.5 by double induction on k and $w(I)$. We first study the case $k = 0$. If $w(I) = 0$, then we have nothing to prove. Assume that Lemma 3.5.5 is true for the cases $k = 0$ and $w(I) \leq w$, so that there is a polynomial $P_w(u_1, \dots, u_w)$ such that

$$\frac{f_0^{(w)}}{f_0} = P_w\left(\frac{f_0'}{f_0}, \left(\frac{f_0'}{f_0}\right)', \dots, \left(\frac{f_0'}{f_0}\right)^{(w-1)}\right),$$

where P_w is an isobaric polynomial of weight w if we associate the weight ℓ with each variable u_ℓ. Then,

$$\frac{f_0^{(w+1)}}{f_0} = \left(\frac{f_0^{(w)}}{f_0}\right)' + \frac{f_0'}{f_0}\frac{f_0^{(w)}}{f_0} = \sum_{j=1}^{w} \frac{\partial P_w}{\partial u_j}\left(\frac{f_0'}{f_0}\right)^{(j)} + \frac{f_0'}{f_0}P_w.$$

If we define

$$P_{w+1}(u_1, \dots, , u_{w+1}) = \sum_{j=1}^{w} \frac{\partial P_w}{\partial u_j} u_{j+1} + u_1 P_w(u_1, \dots, u_w),$$

P_{w+1} is isobaric of weight $w+1$ and we have

$$\frac{f_0^{(w+1)}}{f_0} = P_{w+1}\left(\frac{f_0'}{f_0}, \left(\frac{f_0'}{f_0}\right)' \dots, \left(\frac{f_0'}{f_0}\right)^{(w)}\right).$$

This shows that Lemma 3.5.5 holds for the case $k = 0$ and $w(I) = w + 1$. Therefore, Lemma 3.5.5 for the case $k = 0$ is proved.

We shall next prove Lemma 3.5.5 under the assumption that it is true for the cases $< k$. If $w(I) = 0$, the proof is trivial because we have necessarily $I = I_{0,k}$ in this case. Assume that Lemma 3.5.5 is true for the cases $w(I) < w$ and consider the case $w(I) = w$.

We first study the case $I = (i_0, \dots, i_{k-1}, i_k) \neq I^* := (0, \dots, k-1, k+w)$. We use the notation

$$J(j_0, \dots, j_r; g_0, \dots, g_s) := \begin{pmatrix} g_0^{(j_0)} & g_1^{(j_0)} & \dots & g_s^{(j_0)} \\ & \dots\dots\dots & & \\ g_0^{(j_r)} & g_1^{(j_r)} & \dots & g_s^{(j_r)} \end{pmatrix}$$

for nonnegative integers $j_0, \dots, j_r$ and functions $g_0, \dots, g_r$, and define

$$F := \det \begin{pmatrix} J(0, \dots, k-1; f_0, \dots, f_k) & \mathbf{0} \\ J(i_0, \dots, i_k; f_0, \dots, f_k) & J(i_0, \dots, i_k; f_0, \dots, f_{k-1}) \end{pmatrix},$$

where $\mathbf{0}$ means the (k, k)-matrix whose components are all zero. By the Laplace expansion theorem, we get

$$F = \sum_{\ell=0}^{k} (-1)^{\ell} W(0, \dots, k-1, i_\ell; I_{0,k}) W(i_0, \dots, \hat{i}_\ell, \dots, i_k; I_{0,k-1}),$$

where $\hat{i}_\ell$ means that the index i_ℓ is deleted. On the other hand, by subtracting the $(k + \ell + 1)$-th column from the ℓ-th column for each $\ell = 1, \dots, k$, we obtain

$$F = (-1)^k W(i_0, \dots, i_k; I_{0,k}) W(I_{0,k-1}; I_{0,k-1}).$$

Therefore,

$$\begin{aligned} &\frac{W(i_0, \dots, i_k; I_{0,k})}{W(I_{0,k}; I_{0,k})} \\ &= \sum_{\ell=0}^{k} (-1)^{k+\ell} \frac{W(0, \dots, k-1, i_\ell; I_{0,k})}{W(I_{0,k}; I_{0,k})} \frac{W(i_0, \dots, \hat{i}_\ell, \dots, i_k; I_{0,k-1})}{W(I_{0,k-1}; I_{0,k-1})}. \end{aligned}$$

Since $w(0, \dots, k-1, i_\ell) < w \quad (0 \le \ell \le k)$, $\frac{W(0, \dots, 0, k-1, i_\ell; I_{0,k})}{W(I_{0,k}; I_{0,k})}$ can be written as a polynomial of several functions of the type (3.5.6) which is isobaric of weight $w(0, \dots, k-1, i_\ell) = i_\ell - k$ according to the induction hypothesis on $w(I)$. On the other hand, we can apply the induction hypothesis on k to each function $\frac{W(i_0, \dots, \hat{i}_\ell, \dots, i_k; I_{0,k-1})}{W(I_{0,k-1}; I_{0,k-1})}$. These functions are written as an isobaric polynomial of several functions of the type (3.5.6) whose weight is

$$\begin{aligned} &w(i_0, \dots, \hat{i}_\ell, \dots, i_k) \\ &\qquad = i_0 + \cdots + \hat{i}_\ell + \cdots + i_k - (0 + 1 + \cdots + (k-1)) \\ &\qquad = w(I) - (i_\ell - k). \end{aligned}$$

From these facts, we conclude that $W(i_0, \dots, i_k; I_{0,k}) / W(I_{0,k}; I_{0,k})$ has the desired representation.

It remains to prove Lemma 3.5.5 for the case $(i_0, \dots, i_{k-1}, i_k) = I^*$. As is easily seen by induction on w, we can write

$$\frac{W(I_{0,k}; I_{0,k})^{(w)}}{W(I_{0,k}; I_{0,k})} = \frac{W(0, \dots, k-1, k+w; I_{0,k})}{W(I_{0,k}; I_{0,k})} + \sum_{w(J)=w, J \neq I^*} C_J \frac{W(J; I_{0,k})}{W(I_{0,k}; I_{0,k})}.$$

where the C_J's are constants depending only on J. The left hand side has the desired representation because we can apply the same argument as in the proof for the case $k = 1$ to the function $W(I_{0,k}; I_{0,k})$. On the other hand, as was shown above, the last term of the right hand side also has the desired representation. Accordingly, we obtain the desired representation of $\frac{W(0,\dots,k-1,k+w;I_{0,k})}{W(I_{0,k};I_{0,k})}$. This completes the proof of Lemma 3.5.5.

COROLLARY 3.5.7. *In the same situation as in Lemma 3.5.5, we have*

$$\nu_{W(I;I_{0,k})} \geq \nu_{W(I_{0,k};I_{0,k})} - w(I)$$

for every $I \in \mathcal{I}_k$.

PROOF. The function of the type (3.5.6) has no pole of order larger than ℓ. As a result of Lemma 3.5.5, $W(I; I_{0,k})/W(I_{0,k}; I_{0,k})$ has no pole of order larger than $w(I)$. This gives Corollary 3.5.7.

Let $0 \leq k < n$. We attach labels to all elements in $\mathcal{I}_k^n$ as

$$\mathcal{I}_k^n = \{I_0 := (0, \dots, k), I_1, \dots, I_N\},$$

where $N = N_k := \binom{n+1}{k+1} - 1$. We consider the $(N+1, N+1)$-matrix

$$\mathbf{W} := (W(I_r; I_s); 0 \leq r, s \leq N).$$

As a result of the classical theorem of Sylvester and Franke([67, p. 97]), we have

$$\det(\mathbf{W}) = W(I_{0,n}; I_{0,n})^{\binom{n}{k}} (\not\equiv 0).$$

To study the divisors $\nu_{\det \mathbf{W}}$, we use the following lemma given in [73, p. 41].

LEMMA 3.5.8. *Let f be a nondegenerate holomorphic curve in $P^n(\mathbf{C})$ defined on an open Riemann surface M and p a point in M. If we suitably choose homogeneous coordinates on $P^n(\mathbf{C})$ and a holomorphic local coordinate ζ around p with $\zeta(p) = 0$, then f has a reduced representation $f = (f_0 : \cdots : f_n)$ with holomorphic functions f_i which are expanded as*

$$f_j = \zeta^{\delta_j} + \sum_{i>\delta_j} c_{ji}\zeta^i \quad (c_{ji} \in \mathbf{C})$$

in a neighborhood of p, where

$$\delta_0 := 0 < \delta_1 < \cdots < \delta_n.$$

PROOF. Start with an arbitrary homogeneous coordinates $(w_0 : \cdots : w_n)$ and a holomorphic local coordinate ζ around p with $\zeta(p) = 0$. We shall apply the following three elementary nonsingular linear transformations of homogeneous coordinates:

a) Interchanging of two coordinates w_i and w_j for $i < j$.

b) Multiplication of a coordinate w_i by a nonzero constant.

c) Replacement of a coordinate w_j by $w_j - cw_i$ with an arbitrary constant c for $i < j$.

By the operations a) and b), we can make $f_0(t) = 1 + \cdots$, where "$\ldots$" indicates the sum of terms of higher degrees. By subtracting suitable multiples of f_0 from $f_i(t)$ for $i \geq 1$, we annihilate the constant terms of these n functions. We take the largest positive exponent δ_1 such that the f_i's have representations

$$f_1(t) = t^{\delta_1}\tilde{f}_1(t), \ldots, f_n(t) = t^{\delta_1}\tilde{f}_n(t)$$

with holomorphic functions $\tilde{f}_i$ $(1 \leq i \leq n)$ around p. Here, by the use of the operations a) and b) we can make $\tilde{f}_1(t) = 1 + \cdots$, and by operation c) we can annihilate the constant terms of $\tilde{f}_i (i \geq 2)$. Repeating this processes, we see that Lemma 3.5.8 follows easily.

For the functions $f_j = t^{\delta_j} + \cdots$, we have

$$f_j^{(i)} = \delta_j(\delta_j - 1)\cdots(\delta_j - i + 1)t^{\delta_j - i} + \cdots .$$

Set

$$\phi_i(\delta_j) = \delta_j(\delta_j - 1)\cdots(\delta_j - i + 1).$$

Then, for all $I = (i_0, \dots, i_k)$ and $J = (j_0, \dots, j_k)$ in $\mathcal{I}_k$, we see easily

(3.5.9) $$W(I; J) = A_{IJ}\zeta^{\delta_{j_0}+\cdots+\delta_{j_k}-(i_0+\cdots+i_k)} + \cdots ,$$

where

$$A_{IJ} := \det(\phi_{i_\ell}(\delta_{j_m}); 0 \le \ell, m \le k).$$

The divisor ν_k defined in §3.1 is given by

(3.5.10) $$\nu_k = \nu_{|F_k|} = \min(\nu_{W(I_0,I_0)}, \dots, \nu_{W(I_0,I_N)}).$$

We show here the following:

(3.5.11) *For δ_j's as in Lemma 3.5.8,*

$$\nu_k(z_0) = (\delta_0 - 0) + (\delta_1 - 1) + \cdots + (\delta_k - k).$$

In fact, since $A_{I_0J_0} \neq 0$, we have by (3.5.9)

$$\nu_{W(I_0;I_0)}(z_0) = (\delta_0 - 0) + (\delta_1 - 1) + \cdots + (\delta_k - k)$$

and

$$\nu_{W(I_0;J)}(z_0) > (\delta_0 - 0) + (\delta_1 - 1) + \cdots + (\delta_k - k)$$

for all $J \neq I_0$, which gives (3.5.11).

REMARK 3.5.12. According to (3.5.11), each δ_k can be represented as a linear combination of $\nu_0, \dots, \nu_k$ over $\mathbf{Z}$. This implies that each δ_k does not depend on the choice of a holomorphic local coordinate ζ and homogeneous coordinates on $P^n(\mathbf{C})$ and it defines a divisor on M.

LEMMA 3.5.13. *For all $I, J \in \mathcal{I}_k$, we have*

$$\nu_{W(I;J)} \ge (\nu_k - w(I) + w(J))^+.$$

PROOF. For each point $z_0 \in \mathbf{C}$, we take δ_j's as in Lemma 3.5.8. For $I = (i_0, \dots, i_k)$ and $J = (j_0, \dots, j_k)$ in $\mathcal{I}_k$,

$$\begin{aligned} \nu_{W(I;J)} &\ge (\delta_{j_0} - i_0) + (\delta_{j_1} - i_1) + \cdots + (\delta_{j_k} - i_k) \\ &= \sum_{\ell=1}^{k}(\delta_\ell - \ell) - \sum_{\ell=1}^{k}(i_\ell - \ell) + \sum_{\ell=0}^{k}(\delta_{j_\ell} - \delta_\ell) \\ &\ge \nu_k(z_0) - w(I) + w(J), \end{aligned}$$

because $\delta_{j_\ell} - \delta_\ell = \sum_{m=\ell+1}^{j_\ell}(\delta_m - \delta_{m-1}) \ge j_\ell - \ell$ for $\ell = 0, 1, \dots, k$. Since we always have $\nu_{W(I;J)} \ge 0$, Lemma 3.5.13 holds.

LEMMA 3.5.14 . *Let* f_{rs} $(0 \le r, s \le N)$ *be nonzero holomorphic functions on* $\mathbf{C}$ *with* $\det(f_{rs}) \not\equiv 0$. *Assume that, for nonnegative integers* m *and* w_r $(0 \le r \le N)$,

$$\nu_{f_{rs}} \ge (m - w_r + w_s)^+$$

at a point z_0. *Then,*

$$\nu_{\det(f_{rs})}(z_0) \ge m(N+1).$$

PROOF. By definition, we have

$$\det(f_{rs}) = \sum_{\binom{0 \ \cdots \ N}{i_0 \ \cdots \ i_N}} \operatorname{sgn}\binom{0 \ \cdots \ N}{i_0 \ \cdots \ i_N} F_{i_0 \cdots i_N},$$

where $F_{i_0 \cdots i_N} := f_{0i_0} f_{1i_1} \cdots f_{Ni_N}$. As is easily seen, each $F_{i_0 \cdots i_N}$ satisfies

$$\nu_{F_{i_0 \cdots i_N}} = \sum_{r=0}^{N} \nu_{f_{ri_r}} \ge m(N+1).$$

This concludes Lemma 3.5.14.

As an immediate consequence of Lemmas 3.5.13 and 3.5.14, we have the following:

COROLLARY 3.5.15. $\nu_{\det(\mathbf{W})} \ge \binom{n+1}{k+1} \nu_k.$

Now we shall prove the following inequality for divisors which plays an important role in the next section.

PROPOSITION 3.5.16. *Let* $H_1, H_2, \ldots, H_q$ $(q \ge N+1)$ *be decomposable hyperplanes in* $P^N(\mathbf{C})$ *in general position. Then,*

$$\binom{n}{k} \nu_n \ge \binom{n+1}{k+1} \nu_k + \sum_{j=1}^{q} (\nu(f^k, H_j) - (k+1)(n-k))^+.$$

PROOF. For convenience's sake, let each H_j $(1 \le j \le q)$ be given by

$$H_{j+1} : \ \langle W, A_j^k \rangle = 0 \qquad\qquad j = 0, \ldots, q-1$$

with a decomposable k-vector A_j^k. Take an arbitrary point $z_0 \in \mathbf{C}$. For brevity, we set $m_j := \nu(f^k, H_{j+1})(z_0)$ $(0 \le j \le q-1)$. Changing indices, if necessary, we assume that

$$m_0 \ge m_1 \ge \cdots \ge m_t \ge (k+1)(n-k) > m_{t+1} \ge \cdots \ge m_{q-1}.$$

If $t+1=0$, namely, $(k+1)(n-k) > m_j$ for all j, Proposition 3.5.16 is true because of Corollary 3.5.15. We assume $t \geq 0$. Set

$$F^j = \langle F \wedge F' \wedge \cdots \wedge F^{(k)}, A_j^k \rangle \qquad (j = 0, 1, \ldots, q-1),$$

where $F = (f_0, \ldots, f_n)$. Since $A_0^k, \ldots, A_N^k$ are linearly independent over $\mathbf{C}$, $W(I_0; I_0), \ldots, W(I_0; I_N)$ can be written as linear combinations of $F^0, \ldots, F^N$. If $t \geq N$, then

$$\nu_{W(I_0;I_r)} \geq \min(\nu_{F^0}, \nu_{F^1}, \ldots, \nu_{F^N}) = \nu_k + m_{N+1} > \nu_k$$

for $r = 0, 1, \ldots, N$. This contradicts (3.5.10). Therefore, $t < N$.

Now, we choose $N-t$ vectors $B^{t+1}, \ldots, B^N$ in $\bigwedge^{k+1} \mathbf{C}^{n+1}$ such that $B^0 := A_0^k, \ldots, B^t := A_t^k, B^{t+1}, \ldots, B^N$ are linearly independent, where B^i's are regarded as column vectors. We define the square matrix $\mathbf{B} := (B^0, B^1, \ldots, B^N)$ and

$$\mathbf{U} \equiv (U_r^s; 0 \leq r, s \leq N) \equiv (U^0, U^1, \ldots, U^N) := \mathbf{WB}.$$

Then, since $\det \mathbf{B} \neq 0$, we have $\nu_{\det \mathbf{W}} = \nu_{\det \mathbf{U}}$. Set

$$W_{I_r} := (W(I_r, I_0), W(I_r, I_1), \ldots, W(I_r, I_N)),$$
$$W^{I_s} :=^t (W(I_0, I_s), W(I_1, I_s), \ldots, W(I_N, I_s)).$$

We can write $U_r^s = W_{I_r} B^s$ $(0 \leq r, s \leq N)$ and

$$U^s = \sum_{r=0}^{N} b_r^s W^{I_r} \tag{3.5.17}$$

for $s = 0, 1, \ldots, N$, where $B^s =^t (b_0^s, \ldots, b_N^s)$. By assumption, F^j has a zero of order $m_j + \nu_k(z_0)$ at z_0. We claim that $U_r^s = W_{I_r} A_s^k$ $(s = 0, 1, \ldots, t)$ has a zero of order $\geq \nu_k + m_s - w(I_r)$. To see this, for each s we choose a system of orthonormal bases $E_0, E_1, \ldots, E_n$ of $\mathbf{C}^{n+1}$ such that $A_s^k = cE_0 \wedge E_1 \wedge \ldots \wedge E_k$ for some nonzero constant c. If we take the reduced representation of f with respect to this coordinate system, we can write

$$W_{I_r} A_s^k = cW(i_0, \ldots, i_k; 0, \ldots, k)$$

for each $I_r = (i_0, \ldots, i_k) \in \mathcal{I}_k$. Then, we can apply Corollary 3.5.7 to these functions and our claim is justified.

On the other hand, since $i_\ell - \ell \leq n-k$ $(\ell = 0, 1, \dots, k)$ for all $I = (i_0, \dots, i_k) \in \mathcal{I}_k$, we always have

$$w(I_r) \leq (k+1)(n-k). \tag{3.5.18}$$

Therefore, every component of U^s has a zero of order $\geq m_s - (k+1)(n-k) + \nu_k(z_0)$ at z_0 for $1 \leq s \leq t$. Set

$$\tilde{U}^s := (z - z_0)^{(k+1)(n-k)-m_s} U^s \qquad (s = 0, 1, \dots, t)$$

and $\tilde{\mathbf{U}} := (\tilde{U}^0, \dots, \tilde{U}^t, U^{t+1}, \dots, U^N)$. Then,

$$\det \tilde{\mathbf{U}} = (z - z_0)^{\Sigma_{s=0}^t ((k+1)(n-k)-m_s)} \det \mathbf{U}$$

and, for the r-th component $\tilde{U}^s_r$ of $\tilde{U}^s$ $(0 \leq s \leq t)$, we see that

$$\begin{aligned} \nu_{\tilde{U}^s_r}(z_0) &\geq \nu_k + m_s - w(I_r) + ((k+1)(n-k) - m_s) \\ &= \nu_k - w(I_r) + (k+1)(n-k). \end{aligned} \tag{3.5.19}$$

By virtue of (3.5.17), we can rewrite

$$\det \tilde{\mathbf{U}} = \sum_{I=(i_{t+1},\dots,i_N)\in\mathcal{I}^N_{N-t}} c_I \det(\tilde{U}^0. \dots, \tilde{U}^t, W^{I_{i_{t+1}}}, \dots, W^{I_{i_N}}). \tag{3.5.20}$$

with suitable constants c_I. For each $i_{t+1}, \dots, i_N$, we set

$$G := \det(\tilde{U}^0, \dots, \tilde{U}^t, W^{I_{i_{t+1}}}, \dots, W^{I_{i_N}})$$

We determine the indices $i_0, i_1, \dots, i_t$ $(0 \leq i_0 < \cdots, < i_t \leq N)$ so that $\{i_0, \dots, i_t, i_{t+1}, \dots, i_N\} = \{0, 1, \dots, N\}$. For convenience's sake, we set

$$\tilde{W}^{i_0} := \tilde{U}^0, \dots, \tilde{W}^{i_t} := \tilde{U}^t, \tilde{W}^{i_{t+1}} := W^{I_{i_{t+1}}}, \dots, \tilde{W}^{i_N} := W^{I_{i_N}},$$

and we rewrite

$$G = \operatorname{sgn}\begin{pmatrix} 0 & \cdots & t & t+1 & \cdots & N \\ i_0 & \cdots & i_t & i_{t+1} & \cdots & i_N \end{pmatrix} \det(\tilde{W}^0, \tilde{W}^1, \dots, \tilde{W}^N).$$

For each $s = 0, 1, \dots, N$, the r-the component $\tilde{W}^s_r$ of $\tilde{W}^s$ has a zero of order $\geq \nu_k - w(I_r) + w(I_s)$ at z_0. In fact, if $s \in \{i_{t+1}, \dots, i_n\}$, this is a result of Lemma 3.5.13. On the other hand, if $s = i_{s'}$ for some s' with $0 \leq s' \leq t$, we see that

$$\nu_{\tilde{W}_r^s}(z_0) = \nu_{\tilde{U}_r^{s'}}(z_0) \geq \nu_k - w(I_r) + (k+1)(n-k) \geq \nu_k - w(I_r) + w(I_s)$$

by virtue of (3.5.18) and (3.5.19).

We now apply Lemma 3.5.14 to the matrix $(\tilde{W}^0, \dots, \tilde{W}^N)$ to show that each term on the right hand side of (3.5.20) has a zero of order $\geq (N+1)\nu_k = \binom{n+1}{k+1}\nu_k$ at z_0. Therefore, $\nu_{\det \tilde{U}} \geq \binom{n+1}{k+1}\nu_k$. Consequently,

$$\begin{aligned}
\nu_{\det \mathbf{U}}(z_0) &= \nu_{\det \tilde{\mathbf{U}}}(z_0) + \sum_{j=0}^{t}(m_j - (k+1)(n-k)) \\
&\geq \binom{n+1}{k+1}\nu_k + \sum_{j=0}^{t}(m_j - (k+1)(n-k)) \\
&= \binom{n+1}{k+1}\nu_k + \sum_{j=1}^{q}(\nu(f^k, H_j) - (k+1)(n-k))^+.
\end{aligned}$$

This completes the proof of Proposition 3.5.16.

§3.6 The second main theorem for derived curves

The purpose of this section is to prove the following second main theorem for the derived curves of a nondegenerate holomorphic curve in $P^n(\mathbf{C})$, which was given by J. and H. Weyl ([73]) and improved by the author in [28].

THEOREM 3.6.1. *Let $f : \mathbf{C} \to P^n(\mathbf{C})$ be a nondegenerate holomorphic curve in $P^n(\mathbf{C})$, let $f^k : \mathbf{C} \to P^{N_k}(\mathbf{C})$ be the derived curves of f and let $H_1, \dots, H_q$ be decomposable hyperplanes in $P^{N_k}(\mathbf{C})$ located in general position, where $N_k = \binom{n+1}{k+1} - 1$. Then, for every $\varepsilon > 0$ there exists a set E with $\int_E (1/r)dr < +\infty$ such that, for all $r \notin E$,*

$$(q - N_k - 1 - \varepsilon)T_f^k(r) \leq \sum_{j=1}^{q} N_{f^k}(r, H_j)^{[(k+1)(n-k)]}.$$

For the proof, we need some preliminary considerations.

We shall use the same notation as in the previous section and in §2.6. We first give the following:

LEMMA 3.6.2. $\sum_{h=0}^{k} \sum_{\ell=h}^{n-1+h-k} p_k(\ell, h)\mu_\ell = \binom{n}{k}\nu_n - \binom{n+1}{k+1}\nu_k,$
where μ_k is the divisor of ramification of the k-th derived curve of f defined by (3.2.1) and $p_k(\ell, h)$ is the quantity given by (2.6.13).

PROOF. Since $p_k(\ell, h) = 0$ for $k < h$ or $k - h \geq n - \ell$ by definition, we see

$$\sum_{h=0}^{k} \sum_{\ell=h}^{n-1+h-k} p_k(\ell, h)\mu_\ell = \sum_{h=0}^{k}\sum_{\ell=h}^{n-1} p_k(\ell, h)\mu_\ell$$
$$= \sum_{\ell=0}^{k-1}\left(\sum_{h=0}^{\ell} p_k(\ell, h)\right)\mu_\ell + \sum_{\ell=k}^{n-1}\left(\sum_{h=0}^{k} p_k(\ell, h)\right)\mu_\ell.$$

We use here the formula

$$\binom{n}{k} = \sum_{m\geq 0}\binom{\ell}{k-m}\binom{n-\ell}{m} \qquad (0 \leq \ell \leq m)$$

which follows from the identity $(1+x)^n = (1+x)^\ell(1+x)^{n-\ell}$ for $0 \leq \ell \leq n$. This implies that, for $0 \leq \ell \leq k-1$,

$$\sum_{h=0}^{\ell} p_k(\ell, h) = \sum_{h=0}^{\ell}\sum_{m=k-h}^{k}\binom{n-\ell}{m+1}\binom{\ell+1}{k-m}$$
$$= \sum_{m=k-\ell}^{k}\sum_{h=k-m}^{\ell}\binom{n-\ell}{m+1}\binom{\ell+1}{k-m}$$
$$= \sum_{m=k-\ell}^{k}(\ell-(k-m-1))\binom{n-\ell}{m+1}\binom{\ell+1}{k-m}$$
$$= \sum_{m=k-\ell}^{k}(\ell+1)\binom{n-\ell}{m+1}\binom{\ell}{k-m}$$
$$= (\ell+1)\binom{n}{k+1}$$

and, for $k \le \ell \le n-1$,

$$\begin{aligned}\sum_{h=0}^{k} p_k(\ell,h) &= \sum_{h=0}^{k}\sum_{m=k-h}^{k}\binom{n-\ell}{m+1}\binom{\ell+1}{k-m}\\ &= \sum_{m=0}^{k}(m+1)\binom{n-\ell}{m+1}\binom{\ell+1}{k-m}\\ &= \sum_{m=0}^{k}(n-\ell)\binom{n-\ell-1}{m}\binom{\ell+1}{k-m}\\ &= (n-\ell)\binom{n}{k}.\end{aligned}$$

On the other hand, with the use of Proposition 2.2.11 we have

$$\begin{aligned}\sum_{\ell=0}^{k-1}(\ell+1)\mu_\ell &= \sum_{\ell=0}^{k-1}(\ell+1)(\nu_{\ell-1}-2\nu_\ell+\nu_{\ell+1})\\ &= k\nu_k-(k+1)\nu_{k-1}\end{aligned}$$

and

$$\sum_{\ell=k}^{n-1}(n-\ell)\mu_\ell = \nu_n+(n-k)\nu_{k-1}-(n-k+1)\nu_k.$$

Using these formulas, we obtain the desired identity

$$\begin{aligned}&\sum_{h=0}^{k}\sum_{\ell=h}^{n-1+h-k} p_k(\ell,h)\mu_\ell\\ &\quad= \binom{n}{k+1}(k\nu_k-(k+1)\nu_{k-1})\\ &\qquad+\binom{n}{k}(\nu_n+(n-k)\nu_{k-1}-(n-k+1)\nu_k)\\ &\quad= \binom{n}{k}\nu_n-\binom{n+1}{k+1}\nu_k.\end{aligned}$$

Given decomposable hyperplanes H_j $(1\le j\le q)$ in $P^{N_k}(\mathbf{C})$ located in general position, choose decomposable k-vectors A_j^k such that

$$H_j:\ \langle W, A_j^k\rangle = 0 \qquad (1\le j\le q),$$

where $q > \binom{n+1}{k+1}$. Define a differential form $\omega_{\ell,h}$ and a function $H_{\ell,h}$ by

$$\omega_{\ell,h} \equiv H_{\ell,h} dd^c|z|^2 := = \prod_{j=1}^{q} \left(\frac{\psi_{\ell-1,h-1}(A_j^k)}{\psi_{\ell,h}(A_j^k) \prod_{A^{h-1} \subseteq A^h \subseteq A_j^k} \log^2 \frac{a\varphi_\ell(A^{h-1})}{\varphi_\ell(A^h)}} \right)^{1/p_k(\ell,h)} \Omega_\ell$$

for a sufficiently large a, where $\varphi_\ell(A^h)$ and $\psi_{\ell,h}(A_j^k)$ are the functions defined by (2.6.6) and (2.6.7). We also define Ric $\omega_{\ell,h} := dd^c \log H_{\ell,h}$.

PROPOSITION 3.6.3. *For given $\varepsilon > 0$, there are positive constants $c_{\ell h}$ and a such that, for $k \leq n-1$,*

$$\sum_{h=0}^{k} \sum_{\ell=h}^{n-1+h-k} p_k(\ell,h) \text{Ric } \omega_{\ell h}$$
$$\geq \sum_{h=0}^{k} \sum_{\ell=h}^{n-1+h-k} c_{\ell h}\omega_{\ell,h} + \left(q - \binom{n+1}{k+1}\right)\Omega_k \tag{3.6.4}$$
$$- \varepsilon \sum_{j=1}^{q} \sum_{h=0}^{k} \sum_{\ell=h}^{n-1+h-k} \sum_{A^{h-1} \subseteq A_j^k} \Omega_\ell(A^{h-1}),$$

where $\Omega_\ell(A^{h-1})$ is the form defined by (2.6.6).

PROOF. By definition, the left hand side of (3.6.4) is the sum of three terms I_1, I_2, I_3 defined by

$$I_1 := \sum_{j=1}^{q} \sum_{h=0}^{k} \sum_{\ell=h}^{n-1+h-k} dd^c \log \frac{\psi_{\ell-1,h-1}(A_j^k)}{\psi_{\ell,h}(A_j^k)},$$

$$I_2 := \sum_{j=1}^{q} \sum_{h=0}^{k} \sum_{\ell=h}^{n-1+h-k} dd^c \log \left(\frac{1}{\prod_{A^{h-1} \subseteq A^h \subseteq A_j^k} \log^2 \left(\frac{a\varphi_\ell(A^{h-1})}{\varphi_\ell(A^h)} \right)} \right)$$

$$I_3 := \sum_{h=0}^{k} \sum_{\ell=h}^{n-1+h-k} p_k(\ell,h) \text{ Ric } \Omega_\ell,$$

where Ric $\Omega_\ell := dd^c \log h_\ell$ for $\Omega_\ell = h_\ell dd^c|z|^2$.

We first consider the term I_1. By (2.6.7) we see

$$\psi_{\ell,k}(A_j^k) = \frac{\varphi_\ell(A_j^k)}{\varphi_{\ell+1}(A_j^k)}.$$

Therefore,
(3.6.5)

$$\begin{aligned} I_1 &= \sum_{j=1}^{q} \sum_{\ell=0}^{n-k-1} \sum_{h=0}^{k} dd^c \log \left(\prod_{j=1}^{q} \psi_{\ell+h-1,h-1}(A_j^k) / \prod_{j=1}^{q} \psi_{\ell+h,h}(A_j^k) \right) \\ &= -\sum_{j=1}^{q} \sum_{\ell=k}^{n-1} dd^c \log \left(\prod_{j=1}^{q} \psi_{\ell,k}(A_j^k) \right) \\ &= -dd^c \log \left(\prod_{j=1}^{q} \varphi_k(A_j^k) \right) \\ &= \sum_{j=1}^{q} (\Omega_k - dd^c \log |\langle F_k, A_j^k \rangle|^2) \\ &= q\Omega_k, \end{aligned}$$

where we used the fact that $\langle F_k, A_j^k\rangle$ is a holomorphic function.

To study the term I_2, we use Propositions 2.6.8 and 2.6.16. It follows that, for some positive constant $c_{\ell h}$,

$$\begin{aligned} &\sum_{j=1}^{q} dd^c \log \left(\prod_{A^{h-1} \subseteq A^h \subseteq A_j^k} \frac{1}{\log(a\varphi_\ell(A^{h-1})/\varphi_\ell(A^h))} \right) \\ &\qquad\qquad + \varepsilon \sum_{j=1}^{q} \sum_{A^{h-1} \subseteq A_j^k} \Omega_\ell(A^{h-1}) \\ &\geq c_{\ell h} \sum_{j=1}^{q} \frac{\psi_{\ell-1,h-1}(A_j^k)}{\psi_{\ell,h}(A_j^k)} \left(\prod_{A^{h-1} \subseteq A^h \subseteq A_j^k} \frac{1}{\log^2(a\varphi_\ell(A^{h-1})/\varphi_\ell(A^h))} \right) \Omega_\ell \\ &\geq c_{\ell h} \prod_{j=1}^{q} \left(\frac{\psi_{\ell-1.h-1}(A_j^k)}{\psi_{\ell,h}(A_j^k) \prod_{A^{h-1} \subseteq A^h} \log^2(a\varphi_\ell(A^{h-1})/\varphi_\ell(A^h))} \right)^{1/p_k(\ell,h)} \Omega_\ell. \end{aligned}$$

Therefore, we have

$$I_2 \geq \sum_{h=0}^{k} \sum_{\ell=h}^{n-1+h-k} c_{\ell h}\omega_{\ell,h} - \varepsilon \sum_{j=1}^{q} \sum_{h=0}^{k} \sum_{\ell=h}^{n-1+h-k} \sum_{A^{h-1} \subseteq A_j^k} \Omega_\ell(A^{h-1}).$$

It remains to study the term I_3. According to Proposition 2.2.11,

$$\operatorname{Ric} \Omega_k = \Omega_{k-1} - 2\Omega_k + \Omega_{k+1}.$$

By the same calculation as in the proof of Lemma 3.6.2 we get

$$\sum_{h=0}^{k} \sum_{\ell=h}^{n-1+h-k} p_k(\ell, h) \operatorname{Ric} \Omega_k = \binom{n}{k}\Omega_n - \binom{n+1}{k+1}\Omega_k = -\binom{n+1}{k+1}\Omega_k.$$

Proposition 3.6.3 is an immediate consequence of the above estimates of the terms I_i $(i = 1, 2, 3)$.

Define

$$T_\ell(r; A^h) := \int_0^r \frac{dt}{t} \int_{\Delta_t} \Omega_\ell(A^h).$$

We then have

(3.6.6) $T_\ell(r; A^h) \leq T_f^\ell(r) + O(1).$

For, since $\nu_u \geq 0$ for $u := |\varphi_\ell(A^h)|$ and $\varphi_\ell(A^h) \leq 1$,

$$\begin{aligned} T_\ell(r; A^h) - T_f^\ell(r) &= \int_0^r \frac{dt}{t} \int_{\Delta_t} dd^c \log |\varphi_\ell(A^h)|^2 \\ &\leq \frac{1}{2\pi} \int_0^{2\pi} \log |\varphi_\ell(A^h)(re^{i\theta})| d\theta + O(1) = O(1) \end{aligned}$$

with the use of Proposition 3.1.3. We give next the following Ahlfors' inequality.

PROPOSITION 3.6.7. *In the above situation, for each $\varepsilon > 0$ there are some positive constants a and C such that*

(3.6.8)
$$\begin{aligned} &\int_0^r \frac{dt}{t} \int_{\Delta_t} \frac{\psi_{\ell-1,h}(A_j^k)}{\psi_{\ell,h+1}(A_j^k)} \prod_{A^h \subseteq A^{h+1} \subseteq A_j^k} \frac{1}{\log^2(a\varphi_\ell(A^h)/\varphi_\ell(A^{h+1}))} \Omega_\ell \\ &\leq \varepsilon \sum_{\ell=0}^{n-1} T_f^\ell(r) + C. \end{aligned}$$

PROOF. We apply the formula (3.1.4) to the function

$$u := \prod_{A^h \subseteq A^{h+1} \subseteq A^k} \frac{1}{\log(a\varphi_\ell(A^h)/\varphi_\ell(A^{h+1}))}.$$

It follows that

$$\begin{aligned}
&\int_0^r \frac{dt}{t} \int_{\Delta_t} dd^c \log \prod_{A^h \subseteq A^{h+1} \subseteq A_j^k} \frac{1}{\log^2(a\varphi_\ell(A^h)/\varphi_\ell(A^{h+1}))} \\
&\quad \leq \frac{1}{2\pi} \int_0^{2\pi} \log \prod_{A^h \subseteq A^{h+1} \subseteq A_j^k} \frac{1}{\log(a\varphi_\ell(A^h)/\varphi_\ell(A^{h+1}))} + O(1) \\
&\quad \leq O(1)
\end{aligned}$$

because $\nu_u \geq 0$ and $\varphi_\ell(A^{h+1})/\varphi_\ell(A^h) \leq 1$. Therefore, using Proposition 2.6.8 we obtain

$$\begin{aligned}
&\int_0^r \frac{dt}{t} \int_{\Delta_t} \frac{\psi_{\ell-1,h}(A^k)}{\psi_{\ell,h+1}(A^k)} \left(\prod_{A^h \subseteq A^{h+1} \subseteq A^k} \frac{1}{\log^2(a\varphi_\ell(A^h)/\varphi_\ell(A^{h+1}))} \right) \Omega_\ell \\
&\quad \leq \varepsilon \sum_{A^h \subseteq A_j^k} T_\ell(r; A_h) + C
\end{aligned}$$

for a positive constant C. With the help of (3.6.6) we have Proposition 3.6.7.

We shall next give the following version of the Cowen-Griffiths second main theorem of derived curves.

THEOREM 3.6.9. *In the same situation as in Theorem 3.6.1, for every $\varepsilon > 0$ there exists a set E with $\int_E (1/r)dr < +\infty$ such that, for all $r \notin E$,*

$$\begin{aligned}
&\sum_{h=0}^{k} \sum_{\ell=h}^{n-1+h-k} p_k(\ell, h) N(r, \mu_\ell) + \left(q - \binom{n+1}{k+1} - \varepsilon \right) T_f^k(r) \\
&\qquad \leq \sum_{j=1}^{q} N_{f^k}(r, H_j).
\end{aligned}$$

PROOF. We integrate both sides of (3.6.4) twice to obtain

$$\sum_{h=0}^{k} \sum_{\ell=h}^{n-1+h-k} p_k(\ell, h) \int_0^r \frac{dt}{t} \int_{\Delta_t} \text{Ric } \omega_{\ell h}$$

$$\geq \sum_{h=0}^{k} \sum_{\ell=h}^{n-1+h-k} c_{\ell h} T^{H_{\ell,h}} + \left(q - \binom{n+1}{k+1}\right) T_f^k(r) \tag{3.6.10}$$

$$- \varepsilon \sum_{j=1}^{q} \sum_{h=0}^{k} \sum_{\ell=h}^{n-1+h-k} \sum_{A^{h-1} \subseteq A_j^k} T_\ell(r; A_j^{h-1}),$$

where $T^{H_{\ell,h}}$ denotes the order function of $H_{\ell,h}$ defined by (3.2.3). Here, with the use of (3.1.4) we have

$$p_k(\ell, h) \int_0^r \frac{dt}{t} \int_{\Delta_t} \text{Ric } \omega_{\ell h} + p_k(\ell, h) N(r, \mu_\ell)$$

$$+ \sum_{j=1}^{q} (N_{\ell-1,h-1}(r; A_j^k) - N_{\ell,h}(r; A_j^k)) \tag{3.6.11}$$

$$\leq p_k(\ell, h) \frac{1}{2\pi} \int_0^{2\pi} \log H_{\ell,h}(re^{i\theta}) d\theta + O(1),$$

where $N_{\ell,h}(r; A_j^k)$ denotes the counting function for the zeros of $\psi_{\ell,h}(A_j^k)$. On the other hand, we have

$$\sum_{j=1}^{q} \sum_{h=0}^{k} \sum_{\ell=h}^{n-1+h-k} (N_{\ell-1,h-1}(r; A_j^k) - N_{\ell,h}(r; A_j^k))$$

$$= \sum_{j=1}^{q} \sum_{\ell=k}^{n-1} N_{\ell,k}(r; A_j^k)$$

$$= - \sum_{j=1}^{q} N_k(r, \nu_{\varphi_k(A_j^k)}) = - \sum_{j=1}^{q} N_{f^k}(r, H_j),$$

which is shown by the calculations similar to (3.6.5).

For a given $\varepsilon > 0$ we can take a set E with $\int_E dr/r < \infty$ such that, for each $r \notin E$, the first term of the right hand side of (3.6.11) is bounded by $O(\log T^{H_{\ell,h}}(r)) + O(1)$ according to Proposition 3.2.4. Moreover, the last term of the right hand side of (3.6.10) is bounded by $\varepsilon\left(\sum_{\ell} T_f^{\ell}(r)\right)+O(1)$ because of (3.6.6). This is also bounded by $\varepsilon(O(T_f^k(r)))+O(1)$ according to Corollary 3.2.11. These facts imply that

$$
\begin{aligned}
&\sum_{\ell,h} p_k(\ell,h)N(r,\mu_\ell)+\left(q-\binom{n+1}{k+1}\right)T_f^k(r)\\
&\qquad\le \sum_{\ell,h} p_k(\ell,h)N(r,\mu_\ell)+\sum_{\ell,h} p_k(\ell,h)\int_0^r\frac{dt}{t}\int_{\Delta_t}\operatorname{Ric}\omega_{\ell h}\\
&\qquad\qquad -\delta\sum_{\ell,h} T^{H_{\ell,h}}+\varepsilon\, T_f^k(r)\\
&\qquad\le \sum_{\ell,h}\left(O(\log T^{H_{\ell,h}}(r))-\delta T^{H_{\ell,h}}\right)+\sum_j N_{f^k}(r,H_j)+\varepsilon T_f^k(r)\\
&\qquad\le O(1)+\sum_j N_{f^k}(r,H_j)+\varepsilon T_f^k(r)
\end{aligned}
$$

for some positive constant δ. Thus, we have Theorem 3.6.9.

Now, we are in a position to give the proof of Theorem 3.6.1. Using Proposition 3.5.16 and Lemma 3.6.2, we obtain

$$
\begin{aligned}
&\sum_{j=1}^{q}\min(\nu(f^k,H_j),\ (k+1)(n-k))\\
&\qquad=\sum_{j=1}^{q}\nu(f^k,H_j)-\sum_{j=1}^{q}(\nu(f^k,H_j)-(k+1)(n-k))^+\\
&\qquad\ge\sum_{j=1}^{q}\nu(f^k,H_j)-\left(\binom{n}{k}\nu_n-\binom{n+1}{k+1}\nu_k\right)\\
&\qquad=\sum_{j=1}^{q}\nu(f^k,H_j)-\sum_{h=0}^{k}\sum_{\ell=h}^{n-1+h-k}p_k(\ell,h)\mu_\ell.
\end{aligned}
$$

By the monotonicity of integrals, we have

$$
\begin{aligned}
&\sum_{j=1}^{q}N_{f^k}(r,H_j)^{[(k+1)(n-k)]}\\
&\qquad\ge\sum_{j=1}^{q}N_{f^k}(r,H_j)(r)-\sum_{h=0}^{k}\sum_{\ell=h}^{n-1+h-k}p_k(\ell,h)N(r,\mu_\ell).
\end{aligned}
$$

For every $\varepsilon > 0$ we conclude by Theorem 3.6.9,

$$\sum_{j=1}^{q} N_{f^k}(r, H_j)^{[(k+1)(n-k)]} \geq \left(q - \binom{n+1}{k+1} - \varepsilon\right) T_f^k(r)$$

for all r except on a set of logarithmic measure zero. Thus, Theorem 3.6.1 is completely proved.

Finally, we give the following defect relation for the derived curves.

COROLLARY 3.6.12. *Let $f : \mathbf{C} \to P^n(\mathbf{C})$ be a nondegenerate holomorphic curve and $f^k : \mathbf{C} \to P^{N_k}(\mathbf{C})$ the k-th derived curve of f. Then, for an arbitrarily given decomposable hyperplanes in $P^{N_k}(\mathbf{C})$ in general position,*

$$\sum_{j=1}^{q} \delta_{f^k}(H_j)^{[(k+1)(n-k)]} \leq \binom{n+1}{k+1}.$$

PROOF. This is shown by the same method as in the proof of Theorem 3.3.8 with the use of Theorem 3.6.1.

Since we always have $\delta_{f^k}(H_j)^{[\infty]} \leq \delta_{f^k}(H_j)^{[(k+1)(n-k)]}$, it holds that

$$\sum_{j=1}^{q} \delta_{f^k}(H_j)^{[\infty]} \leq \binom{n+1}{k+1}.$$

This is a result first proved by J. and H. Weyl. On the other hand, in Corollary 3.6.12, we cannot replace the defect $\delta_{f^k}(H_\ell)^{[k(n-k)]}$ by another defect $\delta_{f^k}(H_\ell)^{[M]}$ for a positive number M smaller than $(k+1)(n-k)$. In fact, we can give an example which illustrates this fact. For a concrete example, refer to the original paper [28].

Chapter 4

Modified defect relation for holomorphic curves

§4.1 Some properties of currents on a Riemann surface

We first discuss some elementary properties of currents on an open Riemann surface M. For each nonnegative integer r we denote by $\mathcal{D}_r$ the space of all C^∞ differentiable forms φ of degree r on M whose supports

$$\mathrm{Supp}(\varphi) := \overline{\{a; \varphi(a) \neq 0\}}$$

are compact and, for each $p, q \geq 0$, by $\mathcal{D}_{p,q}$ the set of all $\varphi \in \mathcal{D}_{p+q}$ of type (p, q). For convenience's sake, we set $\mathcal{D}_r := \{0\}$ for $r < 0$ and $\mathcal{D}_{p,q} := \{0\}$ for $p < 0$ or $q < 0$. We denote $\mathcal{D}_r$ or $\mathcal{D}_{p,q}$ simply by $\mathcal{D}$ for the case where the meaning of r or (p, q) is clear. Let $\omega_{p,q}$ be an arbitrarily fixed nowhere zero differential form of type (p, q) on M. Each $\varphi \in \mathcal{D}_r$ has the unique representation

$$\varphi = \sum_{p+q=r} \varphi^{p,q}$$

with $\varphi^{p,q} \in \mathcal{D}_{p,q}$. A sequence $\{\varphi_n\}$ in $\mathcal{D}_r$ is said to converge to φ in $\mathcal{D}_r$ if there is a compact set K such that $\mathrm{Supp}(\varphi_n) \subset K$ for each n and each point $a \in M$ has a neighborhood U satisfying the condition that a holomorphic local coordinate $z = x + \sqrt{-1}y$ is defined on U and, for functions $\tilde{\varphi}_n^{p,q}$ and $\tilde{\varphi}^{p,q}$ with

$$\varphi_n = \sum_{p+q=r} \tilde{\varphi}_n^{p,q} \omega_{p,q}, \quad \varphi = \sum_{p+q=r} \tilde{\varphi}^{p,q} \omega_{p,q},$$

all derivatives $\dfrac{\partial^{k+\ell}}{\partial x^k \partial y^\ell} \tilde{\varphi}_n^{p,q}$ converge to $\dfrac{\partial^{k+\ell}}{\partial x^k \partial y^\ell} \tilde{\varphi}^{p,q}$ uniformly on U.

DEFINITION 4.1.1. We define a *current* T of degree r to be a map of $\mathcal{D}_{2-r}$ into $\mathbf{C}$ such that T is linear over $\mathbf{C}$ and $\lim_{n \to \infty} T(\varphi_n) = T(\varphi)$ for every sequence $\{\varphi_n\}$ which converges to φ in $\mathcal{D}_r$.

Let T be a current on M of degree r. For each (p,q) with $r = p+q$ and $p \geq 0, q \geq 0$, we define a current

$$T_{p,q}(\varphi) := \begin{cases} T(\varphi) & \varphi \in \mathcal{D}_{1-p,1-q} \\ 0 & \text{otherwise,} \end{cases}$$

which we call the (p,q)-component of T. Every current of degree r is uniquely represented as the sum of the (p,q)-components of T for all (p,q) with $r = p+q$ and $p \geq 0, q \geq 0$. A current T is said a (p,q)-current, or to be of type (p,q), if and only if the (p,q)-component of T coincides with T itself.

For a (p,q)-form Ω on M we can define a (p,q)-current $[\Omega]$ by

$$[\Omega](\varphi) = \int_M \Omega \wedge \varphi \qquad (\varphi \in \mathcal{D}_{1-p,1-q}).$$

On the other hand, for a locally integrable function u on M, the $(0,0)$-current $[u]$ is defined by

$$[u](\varphi) = \int_M u\varphi \qquad (\varphi \in \mathcal{D}_{1,1}).$$

They are simply denoted by Ω and u respectively if we have no cause for confusion. As is easily seen, two locally integrable functions u and v define the same currents if and only if $u = v$ almost everywhere on M, which we denote by the abbreviated notation $u = v$ in the following.

Take a divisor ν on M. We can define the $(1,1)$-current $[\nu]$ by

$$[\nu](\varphi) = \sum_{a \in M} \nu(a)\varphi(a) \qquad (\varphi \in \mathcal{D}_{0,0}),$$

where the right hand side is a finite sum because $\mathrm{Supp}(\varphi)$ is compact. The current $[\nu]$ is called the current associated with ν.

For two $(1,1)$-currents T_1 and T_2, by the notation

$$T_1 \leq T_2 \quad \text{or} \quad T_2 \geq T_1$$

we mean that $T_i(\varphi) = \overline{T_i(\bar{\varphi})}$ $(i = 1,2)$ for all $\varphi \in \mathcal{D}_{0,0}$ and $T_1(\varphi) \leq T_2(\varphi)$ for all nonnegative functions $\varphi \in \mathcal{D}$. We say a $(1,1)$-current T is nonnegative if $T \geq 0$. For two C^∞ $(1,1)$-forms Ω_1 and Ω_2 on M, the notation

$$\Omega_1 \leq \Omega_2 \quad \text{or} \quad \Omega_2 \geq \Omega_1$$

means that the Ω's are locally written as $\Omega_i = h_i dd^c|z|^2$ $(i = 1, 2)$ with real-valued functions h_1, h_2 satisfying the condition $h_1 \leq h_2$ for every holomorphic local coordinate z. Obviously, $\Omega_1 \leq \Omega_2$ if and only if $[\Omega_1] \leq [\Omega_2]$.

For later use, we explain here the integral of a nonnegative $(1,1)$-current T on a domain D of M. By C_D we denote the characteristic function of D, the function defined by

$$C_D(p) := \begin{cases} 0 & p \in D \\ 1 & p \notin D. \end{cases}$$

The integral of T over D is defined by

$$(4.1.2) \quad \int_D T := \sup\{T(\varphi); \varphi(\in \mathcal{D}_{0,0}) \text{ is real and } 0 \leq \varphi \leq C_D\} \ (\leq +\infty).$$

Obviously, we have

$$\int_D [\Omega] = \int_D \Omega$$

for every nonnegative $(1,1)$-form Ω of class C^∞.

DEFINITION 4.1.3. For a nonnegative divisor ν on M the *order* of ν is defined by

$$\text{ord}\ (\nu) := \int_M [\nu] \equiv \sum_{z \in M} \nu(z)\ (\leq +\infty).$$

For an arbitrary divisor ν we set $\nu^+ := \max(\nu, 0)$ and $\nu^- := (-\nu)^+$. Then, $\nu := \nu^+ - \nu^-$. Under the assumption that ord $(\nu^+) < +\infty$ or ord $(\nu^-) < +\infty$, we define the *order* of ν by

$$\text{ord}\ (\nu) := \text{ord}\ (\nu^+) - \text{ord}\ (\nu^-).$$

Let T be a current of degree r on M. The differential of T is defined by

$$dT(\varphi) = (-1)^{r-1} T(d\varphi) \qquad (\varphi \in \mathcal{D}_{1-r}),$$

For a (p,q)-current T, ∂T and $\bar{\partial} T$ are the $(p+1,q)$-currents and the $(p,q+1)$-currents defined by

$$\partial T(\varphi) = (-1)^{p+q-1} T(\partial\varphi) \qquad (\varphi \in \mathcal{D}_{-p,1-q}),$$

$$\bar{\partial}T(\varphi) = (-1)^{p+q-1}T(\bar{\partial}\varphi) \qquad (\varphi \in \mathcal{D}_{1-p,-q})$$

respectively. For a general current $T = \bigoplus_{r=p+q} T_{p,q}$ we define

$$\partial T := \bigoplus_{r=p+q} \partial T_{p,q}, \quad \bar{\partial}T := \bigoplus_{r=p+q} \bar{\partial}T_{p,q}.$$

Then, it holds that

$$dT = (\partial + \bar{\partial})T.$$

We define

$$d^cT = \frac{\sqrt{-1}}{4\pi}(\bar{\partial} - \partial)T.$$

As is easily seen from Stokes theorem, we have $d[\Omega] = [d\Omega], \partial[\Omega] = [\partial\Omega]$ and $\bar{\partial}[\Omega] = [\bar{\partial}\Omega]$ for every differential form Ω.

For a function with mild singularities, we can prove the following:

PROPOSITION 4.1.4. *For every function u with mild singularities it holds that*

$$dd^c[\log|u|^2] = [\nu_u] + [dd^c \log|u|^2].$$

PROOF. The formula to be proved is written as

$$\int_M \log|u|^2 dd^c\varphi = \sum_{a\in M} \nu_u(a)\varphi(a) + \int_M \varphi dd^c \log|u|^2$$

for every $\varphi \in \mathcal{D}_0$. Let $\{a_1, \dots, a_k\}$ be the set of all points in $\mathrm{Supp}(\varphi)$ at which $|u|$ is not C^∞ or vanishes. For each a_i we take a holomorphic local coordinate z_i on an open neighborhood U_i of a_i with $z_i(a_i) = 0$. Here, we may assume that $U_i \cap U_j = \emptyset$ for all i, j with $i \neq j$ and that $|u|$ can be written as

$$|u| = |z_i|^{\nu_u(a_i)} u_i^*(z) \prod_j |\log(|z_i|^{m_{ij}} v_{ij}(z_i))|^{\tau_{ij}} \tag{4.1.5}$$

on U_i with some $\tau_{ij} < 0$, nonnegative integers m_{ij} and positive C^∞-functions v_{ij} and u_i^*. Choose a sufficiently small ε such that

$$D_i(\varepsilon) := \{a; |z_i(a)| < \varepsilon\} \subset \overline{D_i(\varepsilon)} \subset U_i \qquad (1 \leq i \leq k).$$

Since

$$d\varphi \wedge d^c\psi = \frac{\sqrt{-1}}{4\pi}(\partial\varphi \wedge \bar{\partial}\psi + \partial\psi \wedge \bar{\partial}\varphi) = d\psi \wedge d^c\varphi$$

for all $\varphi, \psi \in \mathcal{D}$, we have

$$\log |u|^2 dd^c\varphi = d(\log |u|^2 d^c\varphi - \varphi d^c \log |u|^2) + \varphi dd^c \log |u|^2$$

on $\mathrm{Supp}(\varphi) - \{a_1, \dots, a_k\}$. On the other hand, we can easily prove that

$$\lim_{\varepsilon,\varepsilon' \to 0} \int_{\{\varepsilon<|z_i|<\varepsilon'\}} |dd^c \log |\log(|z_i|^{m_{ij}} v_{ij})|| = 0$$

for each i, j, which shows that $dd^c \log |u|^2$ is locally integrable on M as a form whose coefficients are measurable functions. With the use of Stokes theorem, we get

$$\begin{aligned}
\int_M \log |u|^2 dd^c\varphi &= \lim_{\varepsilon\to 0} \int_{M-\cup_i D_i(\varepsilon)} \log |u|^2 dd^c\varphi \\
&= \lim_{\varepsilon\to 0} \int_{\partial(M-(\cup_i D_i(\varepsilon)))} (\log |u|^2 d^c\varphi - \varphi d^c \log |u|^2) + \int_M \varphi dd^c \log |u|^2 \\
&= -\lim_{\varepsilon\to 0} \sum_{i=1}^{k} \int_{\partial D_i(\varepsilon)} (\log |u|^2 d^c\varphi - \varphi d^c \log |u|^2) + [dd^c \log |u|^2](\varphi)
\end{aligned}$$

What we have to prove is that

$$\lim_{\varepsilon\to 0} \int_{\partial D_i(\varepsilon)} \log |u|^2 d^c\varphi - \varphi d^c \log |u|^2 = -\nu_u(a_i)\varphi(a_i)$$

for each i. We may rewrite $|u| = |z_i|^{\nu_g(a_i)}\tilde{u}$ in a neighborhood of $\bar{D}_i(\varepsilon)$ with some function $\tilde{u}$. We then have

$$\begin{aligned}
&\int_{\partial D_i(\varepsilon)} \log |u|^2 d^c\varphi - \varphi d^c \log |u|^2 \\
&\quad = \nu_g(a_i)\left(\int_{\partial D_i(\varepsilon)} \log |z_i|^2 d^c\varphi - \varphi d^c \log |z_i|^2\right) + J(\varepsilon) \\
&\quad = \nu_g(a_i)\left(2\log\varepsilon \int_{\partial D_i(\varepsilon)} d^c\varphi - \frac{1}{2\pi}\int_0^{2\pi} \varepsilon\varphi(\varepsilon e^{i\theta})\frac{d\log r}{dr}\Big|_{r=\varepsilon} d\theta\right) \\
&\qquad + J(\varepsilon) \\
&\quad = \nu_g(a_i)\left(2\log\varepsilon \int_{D_i(\varepsilon)} dd^c\varphi - \frac{1}{2\pi}\int_0^{2\pi} \varphi(\varepsilon e^{i\theta})d\theta\right) + J(\varepsilon)
\end{aligned}$$

because $d^c = \dfrac{1}{4\pi}\varepsilon\dfrac{\partial}{\partial r}d\theta$ on $\partial D_i(\varepsilon)$ by (3.1.6), where

$$J(\varepsilon) := \int_{\partial D_i(\varepsilon)} \log|\tilde{u}|^2 d^c\varphi - \varphi d^c \log|\tilde{u}|^2.$$

As $\varepsilon \to 0$, we can easily show that $J(\varepsilon)$ converges to zero and so the above quantity converges to $-\nu_g(a_i)\varphi(a_i)$. Thus, we obtain Proposition 4.1.4.

As an immediate consequence of Proposition 4.1.4, we have the following Poincaré-Lelong formula.

COROLLARY 4.1.6. *For a nonzero meromorphic function g on a Riemann surface, it holds that*

$$dd^c[\log|g|^2] = [\nu_g].$$

PROOF. This is an immediate consequence of Proposition 4.1.4 because $dd^c \log|g|^2 = 0$ outside a discrete set and so $[dd^c \log|g|^2] = 0$.

REMARK 4.1.7. In terms of currents, the counting function of a divisor ν on $\Delta_{s,R}$ $(0 < s < R)$ is rephrased as

$$N(r,\nu) = \int_s^r \frac{dt}{t}\int_{\Delta_s^t}[\nu], \tag{4.1.8}$$

As in Proposition 3.1.3, we consider a function with mild singularities on $\Delta_{s,R}$, where we assume that $\nu_u(0) = 0$ for the case $s = 0$. As a result of Proposition 4.1.4, we have

$$\int_s^t \frac{dt}{t}\int_{\Delta_{s,t}} dd^c[\log|u|^2] = \int_s^t \frac{dt}{t}\int_{\Delta_{s,t}} dd^c \log|u|^2 + N(r,\nu_u).$$

With this formula, we can combine two terms of the left hand side of (3.1.4) together into one term.

For later use, we show the following proposition, which is a special case of Weyl's lemma ([16, p. 194]).

PROPOSITION 4.1.9. *Let u be a nonnegative real-valued function with mild singularities on M. If it satisfies the equation $dd^c[\log u] = 0$ as a current, then $\log u$ is a harmonic function on M.*

PROOF. The problem is local. For each $a \in M$ take a holomorphic local coordinate z on some neighborhood U of a with $z(a) = 0$. We may assume that $U = \{z; |z| < \varepsilon_0\}$, where $0 < \varepsilon_0 < 1$, and that u is written as

$$u(z) = |z|^{\nu_u(a)} u^* \prod_j \left| \log \left| \frac{1}{g_j(z) v_j(z)} \right| \right|^{\tau_j}$$

on U, where $\tau_j \leq 0$, u^* and v_j are positive C^∞-functions and g_j is a nonzero holomorphic function with $g_j(0) = 0$. Then, on the domains $D^*_\varepsilon := \{z; \varepsilon < |z| < \varepsilon_0\}(0 < \varepsilon < \varepsilon_0)$, we see

$$dd^c \log u^* + \sum_j \tau_j dd^c \log \left| \log \left| \frac{1}{g_j v_j} \right| \right| = dd^c \log u = 0.$$

It then follows that

$$dd^c \log u^* = \sum_j (-\tau_j) \frac{1}{\log^2 |g_j(z) v_j(z)|^2} \frac{|g_j'(z)|^2}{|g_j(z)|^2} (1 + o(1)) dd^c |z|^2.$$

If $\tau_j < 0$ for some j, then the right hand side diverges as z tends to 0, whereas the left hand side converges. Therefore, $\tau_j = 0$ and $\log u^*$ is harmonic. Moreover, by Proposition 4.1.4 we have $\nu(a)[\nu_z] = dd^c[\log |u|^2] = 0$, whence $\nu(a) = 0$. Thus, $\log u = \log u^*$ is harmonic on U. The proof of Proposition 4.1.9 is completed.

§4.2 Metrics with negative curvature

Let M be an open Riemann surface. We study some properties of metrics on M possibly with singularities in discrete subsets of M. We first give the following:

DEFINITION 4.2.1. We call ds^2 a (conformal) *pseudo-metric* on M if for each $a \in M$ we may write $ds^2 = \lambda^2 |dz|^2$ around a with a nonnegative real-valued function λ with mild singularities and a holomorphic local coordinate z.

We say a function u with mild singularities on a domain D is continuous if u has a finite-valued continuous extension to the totality of D. A continuous pseudo-metric ds^2 means a pseudo-metric which is locally written as $ds^2 = \lambda^2 |dz|^2$ with a continuous nonnegative function λ which has mild singularities.

Let ds^2 be a continuous pseudo-metric on an open Riemann surface M. For a piecewise smooth curve γ on M, we can define the length of γ by

$$L_{ds}(\gamma) := \int_\gamma ds.$$

We call M complete with respect to ds^2 if $L_{ds}(\gamma) = \infty$ for any piecewise smooth divergent curve γ on M as in the case where ds^2 is a nowhere vanishing smooth metric.

For a pseudo-metric ds^2 on M which is locally represented by $ds^2 = \lambda^2|dz|^2$ we define the divisor of ds^2 by

$$\nu_{ds} = \nu_\lambda.$$

DEFINITION 4.2.2. For a pseudo-metric $ds^2 = \lambda_z^2|dz|^2$ on M we define the *Ricci form* by

$$\mathrm{Ric}[ds^2] = -dd^c[\log \lambda_z^2]$$

as a current, which is well-defined on the totality of M because it does not depend on the choice of a holomorphic local coordinate z.

The Gaussian curvature K_{ds^2} of a continuous pseudo-metric $ds^2 = \lambda_z^2|dz|^2$ is given by (1.2.10) only on the set $\{ds^2 \neq 0\}$. The Gaussian curvature K_{ds^2} is said to be *strictly negative* if there exists a positive constant C such that $\mathrm{Ric}[ds^2] \leq -C\Omega_{ds^2}$ as currents, where Ω_{ds^2} denotes the area form associated with ds^2, namely, $\Omega_{ds^2} = (\sqrt{-1}/2)\lambda^2 dz \wedge d\bar{z}$ for $ds^2 = \lambda^2|dz|^2$.

For an open disc $\Delta_R := \{z; |z| < R\}$ the metric

$$d\sigma_{\Delta_R}^2 := \left(\frac{2R}{R^2 - |z|^2}\right)^2 |dz|^2$$

is called the Poincaré metric, which is complete and invariant under any holomorphic automorphism of Δ_R and whose curvature is a constant -1. If a Riemann surface M has the universal covering surface $\pi : \tilde{M} \to M$ which is biholomorphic with the unit disc in $\mathbf{C}$, then M carries the metric $d\sigma_M^2$ such that $\pi^*(d\sigma_M^2)$ is the Poincaré metric of $\tilde{M}$, which we call the Pioncaré metric of M. According to the following restatement of the Ahlfors-Schwarz's lemma, this is the unique maximal conformal metric whose Gaussian curvature is a constant -1.

PROPOSITION 4.2.3. *Let $ds^2 = \lambda^2|dz|^2$ be a continuous pseudo-metric on an open Riemann surface M whose universal covering surface is biholomorphic with the unit disc in $\mathbf{C}$. If the Gaussian curvature of ds^2 is strictly negative, then, for some positive constant C depending only on the curvature, it holds that*

$$ds^2 \leq C d\sigma_M^2.$$

PROOF. Replacing M by the universal covering surface, we may assume $M = \Delta_R$. Set $ds^2 = \mu^2|dz|^2$. By assumption, the function μ satisfies the inequality

$$\Delta \log \mu \geq C_1 \mu^2$$

for some positive constant C_1. Dividing μ by $C_1^{1/2}$, we may assume that $C_1 = 1$. By Proposition 1.4.13, we have

$$\mu(z) \leq \frac{2R}{R^2 - |z|^2}.$$

This shows that $ds^2 \leq d\sigma_{\Delta_R}^2$ and hence gives Proposition 4.2.3.

COROLLARY 4.2.4. *There exists no continuous pseudo-metric on the complex plane $\mathbf{C}$ which has strictly negative Gaussian curvature.*

PROOF. Suppose that there is a continuous pseudo-metric $ds^2 = \lambda^2|dz|^2$ on $\mathbf{C}$ with strictly negative Gaussian curvature. Without loss of generality, we may assume that λ has no singularity at the origin. Taking an arbitrary $R > 0$, we consider the restriction of ds^2 to Δ_R. By Proposition 4.2.3, we get

$$\lambda(z) \leq C \frac{2R}{R^2 - |z|^2},$$

where C is a positive constant not depending on R. Fixing a point z and letting R tend to $+\infty$, we have necessarily that $\lambda \equiv 0$, which is a contradiction. Therefore, we conclude Corollary 4.2.4.

To state the main theorem of this section, we introduce a new notation for currents. For two $(1,1)$-currents Ω_1, Ω_2 on some open set U in M and a positive constant c, by the notation

$$\Omega_1 \prec_c \Omega_2$$

we mean that there are a divisor ν and a continuous real-valued bounded function k with mild singularities on U such that $\nu(z) > c$ for each $z \in \mathrm{Supp}(\nu)$ and

$$\Omega_1 + [\nu] = \Omega_2 + dd^c[\log |k|^2]$$

on U. For brevity, by $\Omega_1 \prec \Omega_2$ or $\Omega_2 \succ \Omega_1$ we mean $\Omega_1 \prec_c \Omega_2$ for some positive constant c. Obviously, we have the following:

PROPOSITION 4.2.5. (i) *If $c \geq d$ and $\Omega_1 \prec_c \Omega_2$, then $\Omega_1 \prec_d \Omega_2$.*

(ii) *If $\Omega_1 \prec_c \Omega_2$ and $\Omega_2 \prec_c \Omega_3$ for a positive constant c, then $\Omega_1 \prec_c \Omega_3$.*

(iii) *If $\Omega_1 \prec \Omega_2$, then $a\Omega_1 + \Omega \prec a\Omega_2 + \Omega$ for every Ω and $a > 0$.*

The main purpose of this section is to prove the following:

THEOREM 4.2.6. *Let M be an open Riemann surface with a complete continuous pseudo-metric ds^2 and let $d\tau^2$ be a continuous pseudo-metric on M whose curvature is strictly negative outside a compact set K. Assume that there exists a constant p with $0 < p < 1$ such that*

$$-\mathrm{Ric}[ds^2] \prec_{1-p} p(-\mathrm{Ric}[d\tau^2])$$

on $M - K$. Then, M is of finite type, namely M is biholomorphic with a compact Riemann surface with finitely many points removed.

For the proof of Theorem 4.2.6, we need the following result first given by A. Huber.

THEOREM 4.2.7. *For an open Riemann surface M, if there is a complete metric $d\rho^2$ on M such that*

$$\int_M \max(-K_{d\rho^2}, 0)\Omega_{d\rho^2} < +\infty,$$

then M is of finite type.

The proof is omitted. For the details, refer to [49, Theorem 13, p. 61, Theorem 15, p. 71], [74, Theorem 1, p. 36] and [12].

PROOF OF THEOREM 4.2.6. By assumption, there is a divisor ν and a continuous real-valued bounded function k on $M - K$ with mild singularities such that $\nu(z) > 1 - p$ for every $z \in \mathrm{Supp}(\nu)$ and

$$-\mathrm{Ric}[ds^2] + [\nu] = p(-\mathrm{Ric}[d\tau^2]) + dd^c[\log k^2].$$

Here, we may assume that the divisor ν and the continuous function k with mild singularities are defined on the totality of M and $0 \leq k \leq 1$, and moreover ds^2 and $d\tau^2$ have no zero on M after modifying them on a neighborhood of K. Choose a nowhere zero holomorphic 1-form ω and write

$$ds^2 = \lambda^2|\omega|^2, \quad d\tau^2 = \eta^2|\omega|^2,$$

where λ, η are continuous functions with mild singularities on M. Set

$$u := \frac{k\eta^p}{\lambda}.$$

Then, $v := \log u$ is harmonic on $M - (K \cup \mathrm{Supp}(\nu))$, $\nu_u > 1 - p$ on $\mathrm{Supp}(\nu) - K$ and

$$\lambda = e^{-v}k\eta^p \leq e^{-v}\eta^p. \tag{4.2.8}$$

Define a new pseudo-metric

$$d\rho^2 := e^{-\dfrac{2v}{1-p}}|\omega|^2$$

on M and set $M_1 := \{a \in M; \nu_{d\rho}(a) \geq 0\}$. Then, $d\rho^2$ is a metric on M_1 which is flat on $M_1 - K$ because of Proposition 4.1.9 and $\nu_{d\rho} < -1$ on $M - (K \cup M_1)$ because $\nu_u > 1 - p$ on $\mathrm{Supp}(\nu_u) - K$. For the proof of Theorem 4.2.6, it suffices to show that

(4.2.9) *The surface M_1 is complete with respect to the metric $d\rho^2$.*

In fact, if (4.2.9) is true, then M_1 is of finite type by Theorem 4.2.7 because

$$\int_{M_1} \max(-K_{d\rho^2}, 0)\Omega_{d\rho^2} = \int_K \max(-K_{d\rho^2}, 0)\Omega_{d\rho^2} < \infty.$$

This implies that M is also of finite type.

The proof of (4.2.9) is given by reduction to absurdity. Assume that (4.2.9) is not true. Then

$$d_0 := \mathrm{dist}_{d\rho}(K, \partial M_1) < +\infty,$$

where $\mathrm{dist}_{d\rho}(K, \partial M_1)$ denotes the distance between K and ∂M_1 with respect to the metric $d\rho^2$. Therefore, for a sufficiently small positive number

δ there is a rectifiable curve $\gamma_0(t)$ $(0 \leq t < 1)$ such that $\gamma_0(0) \in K$, $\gamma_0(t)$ tends to ∂M_1 as $t \to 1$ and $L_{d\rho}(\gamma_0) < d_0 + \delta$, where $L_{d\rho}(\gamma_0)$ denotes the length of the curve γ_0 with respect to the metric $d\rho^2$. If we take some t_0 sufficiently near to 1, the point $p_0 := \gamma_0(t_0)$ satisfies the inequalities

$$\mathrm{dist}_{d\rho}(K, p_0) > \frac{d_0}{2}, \quad L_{d\rho}(\gamma_0|[t_0, 1)) < \frac{d_0}{2},$$

where $\gamma|[\alpha, \beta)$ denotes the part of γ from $t = \alpha$ to $t = \beta$.

By assumption, $d\rho^2$ is flat on $M_1 - K$. By Lemma 1.6.7, there is a differentiable map Φ of a disc $\Delta_R := \{w \in \mathbf{C}; |w| < R\}$ onto an open neighborhood U of p_0 in $M_1 - K$ which is a local isometry with respect to the standard metrics on Δ_R and $d\rho^2$ on U, and for a point $a_0 \in \partial\Delta_R$ the Φ-image of the line segment

$$\Gamma : w = ta_0 \qquad (0 \leq t < 1)$$

tends to the boundary of $M_1 - K$ as t tends to 1. Then, we see easily that $R \leq L_{d\rho}(\gamma_0|[t_0, 1)) < d_0/2$. Set $\gamma := \Phi \circ \Gamma$. Then, γ cannot tend to the set K nor $M - (K \cup M_1)$. In fact, if γ tends to K, then we have an absurd conclusion

$$R \geq \mathrm{dist}_{d\rho}(K, p_0) > d_0/2$$

and, if γ tends to the set $M - (K \cup M_1)$, then we have an absurd conclusion $R = L_{d\rho}(\gamma) = +\infty$ because $\nu_{d\rho} < -1$ on $M - (K \cup M_1)$. Therefore, γ tends to the boundary of M.

Now, we shall estimate the length of γ with respect to the metric ds^2. To this end, we define the function $\tilde{\eta}$ by the identity $d\tau = \tilde{\eta} d\rho$. Then, since

$$d\tau = \eta|\omega| = \tilde{\eta} e^{-\frac{v}{1-p}} |\omega|,$$

we have

$$\eta = \tilde{\eta} e^{-\frac{v}{1-p}}.$$

By the use of (4.2.8), the length of γ with respect to ds^2 is estimated by

$$\begin{aligned} L_{ds}(\gamma) &\leq \int_\gamma e^{-v} \eta^p |\omega| = \int_\gamma e^{-v} \tilde{\eta}^p e^{-\frac{pv}{1-p}} |\omega| \\ &= \int_\gamma \tilde{\eta}^p d\rho = \int_\Gamma (\tilde{\eta} \circ \Phi)^p \Phi^*(d\rho) = \int_\Gamma (\tilde{\eta} \circ \Phi)^p |dw|. \end{aligned}$$

On the other hand, by the assumptions on $d\tau$ the curvature of $\Phi^*(d\tau)$ is strictly negative on Δ_R. It follows from Proposition 4.2.3 that

$$\Phi^*(d\tau) = (\tilde{\eta}\circ\Phi)|dw| \leq C_0\, d\sigma_{\Delta_R}$$

for a positive constant C_0. Since the Poincaré metric $d\sigma^2_{\Delta_R}$ is given by

$$d\sigma^2_{\Delta_R} = \left(\frac{2R}{R^2 - |w|^2}\right)^2 |dw|^2,$$

we see

$$\begin{aligned} L_{ds}(\gamma) &\leq C_1 \int_\Gamma \left(\frac{R}{R^2 - |w|^2}\right)^p |dw| \\ &< \frac{C_2}{1-p} R^{1-p} < \infty, \end{aligned}$$

where each C_i denotes some positive constant. This contradicts the completeness of M with respect to ds^2. Therefore, we have (4.2.9) and so the proof of Theorem 4.2.6 is completed.

§4.3 Modified defect relations for holomorphic curves

Let f be a holomorphic map of an open Riemann surface M into $P^n(\mathbf{C})$ and let

$$H : \langle W, A\rangle = 0$$

be a hyperplane in $P^n(\mathbf{C})$ with $f(M) \not\subseteq H$, where A is a unit vector in $\mathbf{C}$. Take a reduced representation $f = (f_0 : \cdots : f_n)$ on M and set $F := (f_0, \ldots, f_n)$. As in Chapter 3, we set $\nu(f, H) = \nu_{F(H)}$ for the function $F(H) := \langle F, A\rangle$.

For a positive integer m or $m = \infty$, we define the m-truncated pull-back of the divisor corresponding to H, considered as a current, by

$$f^*(H)^{[n]} := [\min(\nu(f, H), n)].$$

For brevity, we write $f^*(H) = f^*(H)^{[\infty]}$.

We consider the following condition for a nonnegative number η.

CONDITION 4.3.1. There exists a compact subset K such that

$$f^*(H)^{[n]} \prec \eta\, [\Omega_f] \quad \text{on } M - K. \tag{4.3.2}$$

The constant $\eta := 1$ always satisfies Condition 4.3.1. In fact, if we take a nonzero holomorphic function g with

$$\nu_g = \min(\nu(f, H), n)$$

and set

$$u := \left| \frac{F(H)}{g} \right|, \quad k := \frac{|F(H)|}{|F|},$$

then we see $0 \leq k \leq 1$ and get the identity

$$f^*(H)^{[n]} + dd^c[\log u^2] = dd^c \log |F|^2 + dd^c[\log k^2]$$

as currents. As $dd^c[\log u^2] = [\nu_u]$ by Corollary 4.1.6 and $\nu_u \geq 1$ on $\mathrm{Supp}(\nu_u)$, we have (4.3.2) for $\eta = 1$ and $K = \emptyset$.

DEFINITION 4.3.3. We define the *modified defect* of H for f by

$$D_f(H) := 1 - \inf\{\eta; \eta \text{ satisfies Condition 4.3.1}\}.$$

For convenience' sake, we set $D_f(H) = 0$ if $f(M) \subseteq H$.

The modified defect has the following properties.

PROPOSITION 4.3.4. (i) $0 \leq D_f(H) \leq 1$.

(ii) *If there exists a compact set K and a bounded nonzero holomorphic function g on $M - K$ such that $\nu_g \geq \min(\nu(f, H), n)$ on $M - K$, or particularly, if $\# f^{-1}(H)$ is finite, then $D_f(H) = 1$.*

(iii) *If $\nu(f, H)(a) \geq m$ at every $a \in f^{-1}(H) - K$ for some compact set K, then $D_f(H) \geq 1 - n/m$.*

PROOF. The property (i) is obvious. Assume that there is a bounded nonzero holomorphic function g satisfying the condition in (ii). Then the function $k := |g|$ is a bounded continuous function with mild singularities. Consider the divisor $\nu := \nu_g - \min(\nu(f, H), n)$. Since

$$f^*(H)^{[n]} + [\nu] = \eta \, \Omega_f + dd^c[\log k^2]$$

on $M - K$ for $\eta = 0$ and $\nu \geq 1$ on $\mathrm{Supp}(\nu)$, we have the assertion (ii). To see (iii), we consider the bounded function

$$k = \left(\frac{|F(H)|}{|F|} \right)^{n/m}.$$

It is a continuous function with mild singularities and satisfies the identity

$$f^*(H)^{[n]} + \left[\frac{n}{m}\,\nu(f,H) - \min(\nu(f,H),n)\right] = \frac{n}{m}\,dd^c \log|F|^2 + dd^c[\log k^2]$$

on $M - K$. Since $\nu(f,H) \geq m$ on $f^{-1}(H) - K$, we have

$$\frac{n}{m}\,\nu(f,H) \geq n \geq \min(\nu(f,H),n)$$

on $f^{-1}(H) - K$ and so $f^*(H)^{[n]} \prec (n/m)\,\Omega_f$. Then, we conclude $D_f(H) \geq 1 - n/m$.

Consider the case where f is nondegenerate and M contains an open set G with compact boundary ∂G such that there is a biholomorphic map Φ of a neighborhood D of $M - G$ onto a neighborhood of $\Delta_{s,+\infty} = \{z; s \leq |z| < +\infty\}$ with $\Phi(\partial G) = \{z; |z| = s\}$. For simplicity, we identify D with a neighborhood of $\Delta_{s,+\infty}$ and denote by $\tilde{f}$ the restriction of the given nondegenerate holomorphic map $f : M \to P^n(\mathbf{C})$ to D. For the particular case $s = 0$, after a coordinate change we assume that the given nondegenerate holomorphic map $f : \mathbf{C} \to P^n(\mathbf{C})$ satisfies the condition $f(0) \notin H$. We can prove the following relation between the classical defect and the modified defect.

THEOREM 4.3.5. *If $\tilde{f}$ has an essential singularity at ∞, then*

$$0 \leq D_{\tilde{f}}(H) \leq \delta_{\tilde{f}}(H)^{[n]}.$$

PROOF. Assume that $\tilde{f}$ has an essential singularity at ∞ and choose an arbitrary number η satisfying Condition 4.3.1 for a compact set K. Then there is a nonnegative continuous bounded function k with mild singularities on M such that

$$f^*(H)^{[n]} \leq \eta\Omega_f + dd^c[\log k^2] \tag{4.3.6}$$

on $M - K$. Here we may assume that $0 \leq k \leq 1$. Moreover, changing s if necessary, we may assume that $D \supseteq \Delta_{s,\infty}$ and $\overline{\Delta}_{s,\infty} \cap K = \emptyset$. Set $v := k\,|F|^{\eta}$. By (4.3.6), we have $f^*(H)^{[n]} \leq dd^c[\log v^2]$. Therefore, using

Proposition 3.1.3, Corollary 3.1.12 and Remark 4.1.7, we obtain

$$\begin{aligned}
N_{\tilde{f}}(r,H)^{[n]} &= \int_s^r \frac{dt}{t}\int_{\Delta_{s,t}} f^*(H)^{[n]} \\
&\leq \int_s^r \frac{dt}{t}\int_{\Delta_{s,t}} dd^c[\log v^2] \\
&\leq \frac{1}{2\pi}\int_0^{2\pi} \log|v(re^{i\theta})|d\theta - \frac{1}{2\pi}\int_0^{2\pi} \log|v(se^{i\theta})|d\theta + O(\log r) \\
&\leq \eta\left(\frac{1}{2\pi}\int_0^{2\pi} \log|F|(re^{i\theta})d\theta - \frac{1}{2\pi}\int_0^{2\pi} \log|F|(se^{i\theta})d\theta\right) + O(\log r) \\
&\leq \eta\, T_{\tilde{f}}(r) + O(\log r).
\end{aligned}$$

This implies that

$$\frac{N_{\tilde{f}}(r,H)^{[n]}}{T_{\tilde{f}}(r)} \leq \eta + O\left(\frac{\log r}{T_{\tilde{f}}(r)}\right).$$

On the other hand, since $\tilde{f}$ has an essential singularities at ∞ by assumption, we have

$$\lim_{r\to\infty} \frac{\log r}{T_{\tilde{f}}(r)} = 0.$$

with the use of Proposition 3.3.3. By the definition of the classical defect, as $r \to +\infty$ we have

$$1 - \eta \leq \delta_{\tilde{f}}(H)^{[n]}.$$

Since we can take $1-\eta$ which is arbitrarily near $D_{\tilde{f}}(H)$, we conclude Theorem 4.3.5.

Assume that M is of finite type, namely, biholomorphic with a compact Riemann surface $\bar{M}$ with finitely many points $p_1, \ldots, p_k$ removed. Then each point p_j has a neighborhood U_j such that $U_j = \{p; |z_j(p)| < 1\}$ with $z_j(p_j) = 0$ for a holomorphic local coordinate z_j. With a new local coordinate $\tilde{z}_j = 1/z_j$ we identity U_j^* with $\Delta_{s,+\infty}$ for the set $U_j^* := U_j - \{p_j\}$.

DEFINITION 4.3.7. In the above situation, a holomorphic map f of M into $P^n(\mathbf{C})$ is said to be *transcendental* if f has no holomorphic extension to $\bar{M}$, namely, the restriction to U_j^* has an essential singularity at ∞ for some j.

We easily see by Theorems 4.3.5 and 3.3.8 the following:

COROLLARY 4.3.8. *Let M be an open Riemann surface of finite type and f a nondegenerate transcendental holomorphic map of M into $P^n(\mathbf{C})$. Take $q(> 2N - n + 1)$ hyperplanes $H_j (1 \leq j \leq q)$ in $P^n(\mathbf{C})$ located in N-subgeneral position and choose Nochka weights $\omega(j)$. Then*

$$\sum_{j=1}^{q} \omega(j) D_f(H_j) \leq n + 1.$$

Now, consider an open Riemann surface with a continuous pseudo-metric ds^2 and a holomorphic map

$$f = (f^1, \ldots, f^L) : M \to P^{n_1 n_2 \ldots n_L}(\mathbf{C}),$$

into the so-called Osgood's space

$$P^{n_1 n_2 \ldots n_L}(\mathbf{C}) := P^{n_1}(\mathbf{C}) \times \cdots \times P^{n_L}(\mathbf{C}).$$

We say that f is *nondegenerate* if each component $f^\ell : M \to P^{n_\ell}(\mathbf{C})$ is nondegenerate. To state the modified defect relation, we give another definition.

DEFINITION 4.3.9. A nondegenerate holomorphic map

$$f = (f^1, \ldots, f^L) : M \to P^{n_1 n_2 \ldots n_L}(\mathbf{C})$$

is said to satisfy the condition $(C)_{\rho_1 \cdots \rho_L}$ for nonnegative constants $\rho_1, \ldots, \rho_L$ if there exists a compact set K such that

$$-\mathrm{Ric}[ds^2] \prec \rho_1 \Omega_1 + \cdots + \rho_L \Omega_L$$

on $M - K$, where each Ω_ℓ denotes the pull-back of the Fubini-Study metric form on $P^{n_\ell}(\mathbf{C})$.

Now, we state the following modified defect relation for a holomorphic curve in $P^{n_1 \cdots n_L}(\mathbf{C})$, which will be proved in the next section.

THEOREM 4.3.10. *Let M be an open Riemann surface with a complete continuous pseudo-metric ds^2 and let $f = (f^1, \ldots, f^L)$ be a nondegenerate holomorphic map of M into $P^{n_1 \cdots n_L}(\mathbf{C})$ which satisfies the condition $(C)_{\rho_1 \cdots \rho_L}$. Assume that M is not of finite type or else that some $f^\ell (1 \leq \ell \leq L)$ is transcendental. For each ℓ take hyperplanes $H_{\ell 1}, \ldots, H_{\ell q_\ell}$*

in $P^{n_\ell}(\mathbf{C})$ located in N_ℓ-subgeneral position and choose Nochka weights $\omega_\ell(1), \ldots, \omega_\ell(q_\ell)$ for these hyperplanes, where $q_\ell > 2N_\ell - n_\ell + 1$. Set

$$D_\ell := \sum_{j=1}^{q_\ell} \omega_\ell(j) D_{f^\ell}(H_{\ell j}).$$

Then, it holds that

$$\text{(i)} \qquad D_\ell \leq n_\ell + 1$$

for some ℓ, or else

$$\text{(ii)} \qquad \sum_{\ell=1}^{L} \frac{n_\ell(n_\ell+1)}{2} \frac{\rho_\ell}{D_\ell - n_\ell - 1} \geq 1.$$

We give here some consequences of Theorem 4.3.10 for the particular case where $L = 1$. To state these, we give the following definition.

DEFINITION 4.3.11. Let f be a nondegenerate holomorphic map of M into $P^n(\mathbf{C})$. For an arbitrary compact set K we set

$$\rho_f^K := \inf\{\rho > 0; -\operatorname{Ric}_{ds^2} \prec \rho\Omega_f \text{ on } M - K\},$$

where we set $\rho_f^K := +\infty$ if there is no ρ with the above property. We define the *order* of f by

$$\rho_f := \inf\{\rho_f^K; K \text{ is an arbitrary compact subset of } M\}.$$

COROLLARY 4.3.12. *Let M be an open Riemann surface with a complete continuous pseudo-metric ds^2 and let f be a nondegenerate holomorphic map of M into $P^n(\mathbf{C})$. Take q ($> 2N - n + 1$) hyperplanes H_j in N-subgeneral position and Nochka weights $\omega(j)$ ($1 \leq j \leq q$) for them. If M is not of finite type or else if f is transcendental, then*

$$\text{(4.3.13)} \qquad \sum_{j=1}^{q} \omega(j) D_f(H_j) \leq n + 1 + \frac{\rho_f n(n+1)}{2}.$$

For the proof, consider the case where $L = 1$ in Theorem 4.3.10.

COROLLARY 4.3.14. *Let M be an open Riemann surface with a complete continuous pseudo-metric ds^2 and let f be a nondegenerate holomorphic map of M into $P^n(\mathbf{C})$. If M is not of finite type or else if f is transcendental, then we have*

$$\sum_{j=1}^{q} D_f(H_j) \leq 2N - n + 1 + \frac{\rho_f n(2N - n + 1)}{2} \tag{4.3.15}$$

for arbitrary hyperplanes $H_1, \ldots, H_q$ in $P^n(\mathbf{C})$ located in N-subgeneral position.

PROOF. Let θ be a Nochka constant for the H_j's. Then, we have

$$\begin{aligned}
\sum_{j=1}^{q} D_f(H_j) &= q + \sum_{j=1}^{q}(D_f(H_j) - 1) \\
&\leq q + \sum_{j=1}^{q} \frac{\omega(j)}{\theta}(D_f(H_j) - 1) \\
&= q + \frac{1}{\theta}\sum_{j=1}^{q} \omega(j) D_f(H_j) - \frac{1}{\theta}\sum_{j=1}^{q}\omega(j) \\
&\leq q + \frac{1}{\theta}(n + 1 + \rho_f \sigma_n) - (q - 2N + n - 1) - \frac{1}{\theta}(n+1) \\
&= 2N - n + 1 + \frac{\rho_f \sigma_n}{\theta} \\
&\leq 2N - n + 1 + \frac{\rho_f n(2N - n + 1)}{2}
\end{aligned}$$

with the help of Theorem 2.4.11, (ii) and (iii), where $\sigma_n := n(n+1)/2$. This gives Corollary 4.3.14.

COROLLARY 4.3.16. *Let M be an open Riemann surface with a complete continuous pseudo-metric ds^2 and let f be a nondegenerate holomorphic map of M into $P^n(\mathbf{C})$. If M is not of finite type or else if f is transcendental, then, for arbitrary hyperplanes $H_1, \ldots, H_q$ in $P^n(\mathbf{C})$ located in general position,*

$$\sum_{j=1}^{q} D_f(H_j) \leq n + 1 + \frac{\rho_f n(n+1)}{2}.$$

PROOF. By definition, general position means n-subgeneral position. Set $N = n$ in Corollary 4.3.14. We then have Corollary 4.3.16.

§4.4 The proof of the modified defect relation

The purpose of this section is to prove Theorem 4.3.10. To this end, we first give the following:

LEMMA 4.4.1. *Let f be a nondegenerate holomorphic map of an open Riemann surface M into $P^n(\mathbf{C})$ and let $H_1, \ldots, H_q$ be hyperplanes in $P^n(\mathbf{C})$ located in N-subgeneral position, where $q > 2N - n + 1$, and take Nochka weights $\omega(j)$ and a Nochka constant θ for these hyperplanes. Assume that*

$$f^*(H_j)^{[n]} \prec \eta_j \Omega_f \qquad (1 \leq j \leq q)$$

on $M - K$ for a compact set K and

$$\gamma := \theta(q - 2N + n - 1) - \sum_{j=1}^{q} \omega(j)\eta_j > \rho\sigma_n$$

for $1 \leq j \leq q$. Set $\sigma_n := n(n+1)/2$ and $\tau_n := \sum_{k=1}^{n} \sigma_k$. For arbitrary positive numbers ρ with $\gamma > \rho\sigma_n$ and δ with $\rho\sigma_n/\gamma < \delta \leq 1$ we can find a positive number ε and a continuous pseudo-metric $d\tau^2$ on M whose curvature is strictly negative on $M - K$ such that

$$\gamma > \varepsilon\sigma_{n+1}, \tag{4.4.2}$$

and

$$0 < p := \frac{\rho(\sigma_n + \varepsilon\tau_n)}{\gamma - \varepsilon\sigma_{n+1}} < \delta \tag{4.4.3}$$

$$\rho\Omega_f \prec_{\delta - p} p(-\mathrm{Ric}[d\tau^2]),$$

where we can choose ε so that $\delta - p$ is arbitrarily small.

PROOF. By assumption, for each j there exist divisors $[\nu_j]$ and bounded continuous functions k_j with mild singularities such that $\nu_j > c_j$ on $\mathrm{Supp}(\nu_j)$ for some positive constants c_j and

$$f^*(H_j)^{[n]} + [\nu_j] = \eta_j \Omega_f + dd^c[\log k_j^2]$$

on $M - K$. Here, we may assume that $0 \leq k_j \leq 1$. Set $h_j := k_j|F|^{\eta_j}$. Then, each $\log h_j$ is harmonic on $M - (K \cup \mathrm{Supp}(\nu_{h_j}) \cup f^{-1}(H_j))$ by the help of Proposition 4.1.9 and $\nu_{h_j} - \min(\nu(f, H_j), n) = \nu_j$.

Take an arbitrary positive number ε with $\gamma > \varepsilon\sigma_{n+1}$ and consider the number p satisfying the condition (4.4.3). For an arbitrary holomorphic local coordinate z, using the same notation as in §2.5, we set

$$\eta_z := \left(\frac{|F|^{\gamma - \varepsilon\sigma_{n+1}} |F_n| \prod_{j=1}^{q} |h_j|^{\omega(j)} \prod_{k=0}^{n} |F_k|^{\varepsilon}}{\prod_{j=1}^{q} (|F(H_j)| \prod_{k=0}^{n-1} \log(a/\varphi_k(H_j)))^{\omega(j)}} \right)^{\frac{1}{\sigma_n + \varepsilon\tau_n}}, \tag{4.4.4}$$

and define the pseudo-metric $d\tau^2 := \eta_z^2 |dz|^2$. Here, we set $\eta_z := 0$ if $\varphi_k(H_j) = 0$. If we choose another local coordinate u instead of z, each $|F_k|$ is multiplied by $\left|\frac{dz}{du}\right|^{k(k+1)/2}$ by Proposition 2.1.6, and so η_z is multiplied by $\left|\frac{dz}{du}\right|$. This means that $d\tau^2$ is well-defined on the totality of $M - K$ independently of the choice of holomorphic local coordinate z. On the other hand, if we set

$$\varphi := \frac{|F_n|}{\prod_{j=1}^{q} |F(H_j)|^{\omega(j)}},$$

then by Lemma 3.2.13

$$\nu^* := \nu_\varphi + \sum_{j=1}^{q} \omega(j) \min(\nu(f, H_j), n) > c$$

on $\mathrm{Supp}(\nu^*)$ for some $c > 0$. Moreover,

$$\nu_0 := \nu_\varphi + \sum_{j=1}^{q} \omega(j) \nu_{h_j} > c' \tag{4.4.5}$$

on $\mathrm{Supp}(\nu_0)$ for some $c' > 0$. In fact,

$$\begin{aligned}
\nu_0 &= \sum_{j=1}^{q} \omega(j) \left(\nu_{h_j} - \min(\nu(f, H_j), n)\right) \\
&\qquad + \left(\nu_\varphi + \sum_{j=1}^{q} \omega(j) \min(\nu(f, H_j), n) \right) \\
&\geq \sum_{j=1}^{q} \omega(j)(\nu_{h_j} - \min(\nu(f, H_j), n)) + c \\
&> \min\{\omega(j)c_j; 1 \leq j \leq q\} + c =: c'
\end{aligned}$$

on $\mathrm{Supp}(\nu_0)$. As a result of (4.4.5), we see $\nu_0 \geq 0$ on $M - K$ and so $d\tau^2$ is a continuous pseudo-metric there. We suitably extend $d\tau^2$ to a continuous pseudo-metric on the totality of M.

We shall next prove that $d\tau^2$ has strictly negative curvature on $M-K$. To this end, we recall Proposition 2.5.7 and Theorem 2.5.3. These yield that

$$\begin{aligned} dd^c \log \eta_z^2 &\geq \frac{\gamma - \varepsilon\sigma_{n+1}}{\sigma_n + \varepsilon\tau_n}\Omega_f + \frac{\varepsilon}{2(\sigma_n + \varepsilon\tau_n)} dd^c \log(|F_0|^2 \cdots |F_{n-1}|^2) \\ &\quad + \frac{1}{2(\sigma_n + \varepsilon\tau_n)} dd^c \log \frac{\prod_{k=0}^{n-1} |F_k|^{2\varepsilon}}{\prod_{1\leq j\leq q, 0\leq k\leq n-1} \log^{4\omega(j)}(a/\varphi_k(H_j))} \\ &\geq \frac{\varepsilon}{2} \frac{\tau_n}{\sigma_n(\sigma_n + \varepsilon\tau_n)} \left(\frac{|F_0|^2 \ldots |F_n|^2}{|F_0|^{2\sigma_{n+1}}} \right)^{1/\tau_n} dd^c|z|^2 \\ &\quad + C_0 \left(\frac{|F_0|^{2\theta(q-2N+n-1)}|F_n|^2}{\prod_{j=1}^q (|F(H_j)|^2 \prod_{k=0}^{n-1} \log^2(a/\varphi_k(H_j)))^{\omega(j)}} \right)^{1/\sigma_n} dd^c|z|^2 \\ &\geq C_1 \left(\frac{|F|^{\theta(q-2N+n-1)-\varepsilon\sigma_{n+1}}|F_n| \prod_{k=0}^n |F_k|^\varepsilon}{\prod_{j=1}^q (|F(H_j)| \prod_{k=0}^{n-1} \log(a/\varphi_k(H_j)))^{\omega(j)}} \right)^{2/(\sigma_n+\varepsilon\tau_n)} dd^c|z|^2 \end{aligned}$$

by the help of Hölder's inequality, where C_0, C_1 are some positive constants. On the other hand, since $|h_j| \leq |F|^{\eta_j}$,

$$\begin{aligned} &|F|^{\theta(q-2N+n-1)-\varepsilon\sigma_{n+1}} \\ &\qquad = |F|^{\gamma-\varepsilon\sigma_{n+1}+\omega(1)\eta_1+\cdots+\omega(q)\eta_q} \geq |F|^{\gamma-\varepsilon\sigma_{n+1}} h_1^{\omega(1)} \cdots h_q^{\omega(q)} \end{aligned}$$

on $M - K$. This concludes that $dd^c \log \eta^2 \geq C_2\eta^2$ for some positive constant C_2. Therefore, $d\tau^2$ has strictly negative curvature on $M - K$.

Now, we represent each hyperplane $H_j (1 \leq j \leq q)$ as

$$H_j : a_{j0}w_0 + \cdots + a_{jn}w_n = 0.$$

For some holomorphic local coordinate z and each pair (j, k) of indices, we choose indices $i_1, \ldots, i_k$ with $1 \leq i_1 < \cdots < i_k \leq q$ such that

$$\psi_{jk}^z := \sum_{\ell \neq i_1, \ldots, i_k} a_{j\ell} W(f_\ell, f_{i_1}, \ldots, f_{i_k}) \not\equiv 0.$$

This is possible because of Remark 2.3.7, (iv). For convenience's sake, we set $\psi^z_{j0} = F(H_j)$. By the theorem of identity, $\psi^z_{jk} \not\equiv 0$ for every holomorphic local coordinate z. We now define

$$k := \left(\frac{\prod_{1 \le j \le q} \prod_{0 \le k \le n-1} |\psi^z_{jk}|^{\varepsilon/q} \log^{\omega(j)}(a/\varphi_k(H_j))}{\prod_{0 \le k \le n-1} |F_k|^{\varepsilon}} \right)^{\frac{1}{\sigma_n + \varepsilon\tau_n}}.$$

It is easily seen that k is a well-defined function on $M - K$ which does not depend on a choice of a holomorphic local coordinate z. Moreover, k is bounded because

$$\begin{aligned} &\frac{|\psi^z_{jk}|^{\varepsilon/q} \log^{\omega(j)}(a/\varphi_k(H_j))}{|F_k|^{\varepsilon/q}} \\ &\qquad \le \left(\frac{|F_k(H_j)|}{|F_k|} \right)^{\varepsilon/q} \log^{\omega(j)}(a/\varphi_k(H_j)) \\ &\qquad \le \sup_{0 < x \le 1} x^{\varepsilon/q} \log^{\omega(j)} \left(\frac{a}{x^2} \right) < +\infty. \end{aligned}$$

Set

$$v_1 := |\varphi| \prod_{j=1}^{q} |h_j|^{\omega(j)}, \quad v_2 := \prod_{j=1}^{q} \prod_{k=0}^{n-1} |\psi^z_{jk}|^{\varepsilon/q}, \quad v := v_1 v_2.$$

Then $\log v$ is harmonic on $M - (K \cup \mathrm{Supp}(\nu_v))$ and $\nu_{v_1} \ge c''$ on $\mathrm{Supp}(\nu_{v_1})$ for some positive c'', not depending on each ϵ, by virtue of (4.4.5). On the other hand, $\nu_{v_2} \ge \varepsilon/q$ on $\mathrm{Supp}(\nu_{v_2})$. We consider the function η_z defined by (4.4.4) for some ε satisfying the condition that

$$\frac{\delta\gamma - \rho\sigma_n}{\delta\sigma_{n+1} + \rho\tau_n} > \varepsilon > \max \left(\frac{\delta\gamma - \rho\sigma_n}{\rho/q + \delta\sigma_{n+1} + \rho\tau_n}, \frac{\delta\gamma - \rho\sigma_n - c''\rho}{\delta\sigma_{n+1} + \rho\tau_n} \right).$$

and, moreover, $\delta - p$ is sufficiently small. Then, we have $0 < p < \delta$ and

$$\nu_v \ge \min\left(\frac{\varepsilon}{q}, c'' \right) > \frac{(\gamma - \varepsilon\sigma_{n+1})(\delta - p)}{\rho}$$

on $\mathrm{Supp}(\nu_v) - K$. On the other hand, it holds that

$$(k\eta_z)^{\sigma_n + \varepsilon\tau_n} = |F|^{\gamma - \varepsilon\sigma_{n+1}} v$$

and so

$$|F|^{\rho} v^{\rho/(\gamma-\varepsilon\sigma_{n+1})} = (k\eta_z)^p.$$

Therefore, we have

$$\rho\Omega_f + \left[\frac{\rho}{\gamma - \varepsilon\sigma_{n+1}}\nu_v\right] = p\, dd^c \log \eta_z^2 + dd^c \log k^{2p}.$$

This gives that

$$\rho\Omega_f \prec_{\delta-p} p(-\mathrm{Ric}_{d\tau^2})$$

on $M - K$. Lemma 4.4.1 is completely proved.

Now, we start to prove Theorem 4.3.10. As in Theorem 4.3.10, let M be an open Riemann surface, let ds^2 be a complete continuous pseudo-metric on M, let $f = (f^1, \dots, f^L)$ be a nondegenerate holomorphic map of M into $P^{n_1 \dots n_L}(\mathbf{C})$, let $H_{\ell j}$ $(1 \le j \le q_\ell)$ be hyperplanes in $P^{n_\ell}(\mathbf{C})$ located in N_ℓ-subgeneral position and let $\omega_\ell(j)$ and θ_ℓ be Nochka weights and a Nochka constant for these hyperplanes respectively, where $q_{n_\ell} \ge 2N_\ell - n_\ell + 1$. Set

$$D_\ell := \sum_{j=1}^{q_\ell} \omega_\ell(j) D_{f^\ell}(H_{\ell j})$$

as in Theorem 4.3.10. If M is of finite type and some f^ℓ is transcendental, then we have $D_\ell \le n_\ell + 1$ by Corollary 4.3.8. Therefore, Theorem 4.3.10 is true in this case. We may assume that M is not of finite type and that $D_\ell > n_\ell + 1$ for each $\ell = 1, \dots, L$. Suppose that the conclusion of Theorem 4.3.10, (ii) does not hold, namely,

$$\sum_{\ell=1}^{L} \frac{\sigma_{n_\ell}\rho_\ell}{D_\ell - n_\ell - 1} < 1.$$

By the assumption and Definition 4.3.3, we can find some positive numbers $\eta_{\ell j}$ $(1 \le j \le q_\ell)$ such that

$$(f^\ell)^*(H_{\ell j})^{[n_\ell]} \prec \eta_{\ell j}\Omega_\ell$$

on $M - K$,

$$\gamma_\ell := \sum_{j=1}^{q_\ell} \omega_\ell(j)(1 - \eta_{\ell j}) - n_\ell - 1 > 0 \qquad (1 \le \ell \le L)$$

and

$$\sum_{\ell=1}^{L} \frac{\sigma_{n_\ell} \rho_\ell}{\gamma_\ell} < 1,$$

where each Ω_ℓ denotes the pull-back of Fubini-Study metric form on $P^{n_\ell}(\mathbf{C})$ by f^ℓ. We note here that the constant γ_ℓ is given by

$$\gamma_\ell = \theta_\ell(q_\ell - 2N_\ell + n_\ell - 1) - \sum_{j=1}^{q_\ell} \omega_\ell(j)\eta_{\ell j}$$

according to Theorem 2.4.11, (ii). On the other hand, by assumption there is a positive constant c_0 such that

$$-\mathrm{Ric}[ds^2] \prec_{c_0} \rho_1\Omega_1 + \cdots + \rho_L\Omega_L. \tag{4.4.6}$$

Using Lemma 4.4.1, we take a positive constant ε_1 and a continuous pseudo-metric $d\tau_1^2$, whose curvature is strictly negative on $M - K$ such that $\gamma_1 > \varepsilon_1 \sigma_{n_1+1}$ and

$$0 < p_1 := \frac{\rho_1(\sigma_{n_1} + \varepsilon_1 \tau_{n_1})}{\gamma_1 - \varepsilon_1 \sigma_{n_1+1}} < 1, \qquad \rho_1\Omega_1 \prec_{1-p_1} p_1(-\mathrm{Ric}[d\tau_1^2]),$$

where we may assume that $1 - p_1 < c_0$. Apply Lemma 4.4.1 again for the constant $\delta = 1 - p_1$ in this case. We can choose a positive constant ε_2 and a continuous pseudo-metric $d\tau_2^2$ with strictly negative curvatures on $M - K$ such that $\gamma_2 > \varepsilon_2 \sigma_{n_2+1}$ and

$$0 < p_2 := \frac{\rho_2(\sigma_{n_2} + \varepsilon_2 \tau_{n_2})}{\gamma_2 - \varepsilon_2 \sigma_{n_2+1}} < 1 - p_1, \qquad \rho_2\Omega_2 \prec_{1-p_1-p_2} p_2(-\mathrm{Ric}\ [d\tau_2^2]).$$

Repeating these processes, we can find positive numbers $p_1, \ldots, p_L$ and continuous pseudo-metrics $d\tau_\ell^2$ with strictly negative curvatures on $M - K$ such that $p := p_1 + \cdots + p_L < 1$, $1 - p < c_0$ and

$$\rho_\ell\Omega_\ell \prec_{1-p_1-\cdots-p_\ell} p_\ell(-\mathrm{Ric}[d\tau_\ell^2])$$

for each $\ell = 1, 2, \ldots, L$. Taking an arbitrary holomorphic local coordinate z, we write

$$d\tau_\ell^2 = \lambda_\ell^2 |dz|^2$$

for each ℓ and define a new continuous pseudo-metric

$$d\tau^2 := (\lambda_1^{p_1} \cdots \lambda_L^{p_L})^{\frac{2}{p}} |dz|^2.$$

Then, we see

$$-\mathrm{Ric}[d\tau^2] = \frac{1}{p} \left(\sum_{\ell=1}^{L} p_\ell (-\mathrm{Ric}[d\tau_\ell^2]) \right). \tag{4.4.7}$$

Therefore, by (4.4.6) and (4.4.7)

$$-\mathrm{Ric}[ds^2] \prec_{1-p} \sum_{\ell=1}^{L} p_\ell (-\mathrm{Ric}[d\tau_\ell^2]) = p\ (-\mathrm{Ric}[d\tau^2]).$$

On the other hand, since the $d\tau_\ell^2$ have strictly negative curvatures,

$$\begin{aligned} dd^c \log(\lambda_1^{p_1} \cdots \lambda_L^{p_L}) &= \sum_{\ell=1}^{L} p_\ell dd^c \log \lambda_\ell \\ &\geq C_0 \sum_{\ell=1}^{L} p_\ell \lambda_\ell^2 dd^c |z|^2 \\ &\geq C_1 \, (\lambda_1^{p_1} \cdots \lambda_L^{p_L})^{\frac{2}{p}} \, dd^c |z|^2 \\ &= C_2 d\tau^2 \end{aligned}$$

for some positive constants C_0, C_1 and C_2. This shows that the Gaussian curvature of $d\tau^2$ is strictly negative on $M - K$. In this situation, we can conclude $p \geq 1$ because p, ds^2 and $d\tau^2$ satisfy all assumptions of Theorem 4.2.6 and M is of finite type. This is a contradiction. The proof of Theorem 4.3.10 is completed.

Chapter 5

The Gauss map of complete minimal surfaces in $\mathbf{R}^m$

§5.1 Complete minimal surfaces of finite total curvature

Let $x = (x_1, \ldots, x_m) : M \to \mathbf{R}^m$ be a minimal surface immersed in $\mathbf{R}^m$ and ds^2 the metric on M induced from $\mathbf{R}^m$. According to (1.2.12) and Definition 4.2.2, the Ricci curvature $\mathrm{Ric}[ds^2]$ is nonpositive. Wc give the following:

DEFINITION 5.1.1. The (normalized) *total curvature* of a minimal surface M is defined by

$$C(M) := \int_M \mathrm{Ric}\ [ds^2] \ \ (\geq -\infty).$$

In this section, we shall explain some properties of complete minimal surfaces of finite total curvature given by S. S. Chern and R. Osserman(cf., [61] and [12]).

We consider the forms

$$\omega_i \equiv f_i dz := \frac{\partial x_i}{\partial z} dz \qquad (1 \leq i \leq m)$$

on M. As explained in §1.2, these are holomorphic and the Gauss map G of M is given by

$$G = (\omega_1 : \cdots : \omega_m) : M \to P^{m-1}(\mathbf{C}). \tag{5.1.2}$$

Now, we give the following characterization of minimal surfaces of finite total curvature.

THEOREM 5.1.3. *For a nonflat complete minimal surface $x : M \to \mathbf{R}^m$ immersed in $\mathbf{R}^m$, the following conditions are equivalent.*

(i) *M has finite total curvature.*

(ii) *There is a compact Riemann surface $\bar{M}$ such that the surface M canonically considered as a Riemann surface is biholomorphic to $\bar{M}$ with finitely many points $p_1, \ldots, p_k$ excluded and each ω_i $(1 \leq i \leq q)$ can be extended to $\bar{M}$ as a meromorphic form.*

(iii) *The surface M is biholomorphic with an open subset of a compact Riemann surface $\bar{M}$ and the Gauss map $G : M \to P^{m-1}(\mathbf{C})$ can be extended to a holomorphic map of $\bar{M}$ into $P^{m-1}(\mathbf{C})$.*

For the proof, we need the following:

LEMMA 5.1.4. *Let h be a nowhere zero holomorphic function on $\Delta_a^* := \{z; 0 < |z| < a\}$ $(0 < a < 1)$ and consider the metric $ds^2 := |h|^2|dz|^2$. If every piecewise smooth continuous curve in Δ_a^* converging to the origin has infinite length, then h has at worst a pole at the origin.*

PROOF. Set $u := \log |h|$. By assumption, u is harmonic on Δ_a^*. For $0 < r < a$, set

$$\beta \equiv \beta_1 + \sqrt{-1}\beta_2 := \frac{1}{2\pi\sqrt{-1}} \int_{|z|=r} \frac{\partial u}{\partial z} \qquad (\beta_1, \beta_2 \in \mathbf{R}),$$

which does not depend on each r because $\partial u/\partial z$ is holomorphic. We define

$$g(z) := 2\int_{z_0}^{z} \left(\frac{\partial u}{\partial z}(\zeta) - \frac{\beta}{\zeta} \right) d\zeta$$

for an arbitrarily fixed $z_0 \in \Delta_a^*$, which is a well-defined single-valued holomorphic function on Δ_a^*. Then,

$$\tilde{u}(z) := \text{Re } g(z) = u(z) - (\beta_1 \log |z| - \beta_2 \arg z) - u(z_0).$$

Since $\tilde{u}$ and u are both single-valued, we have necessarily $\beta_2 = 0$ and so β is real. Here, we use the Laurent expansion

$$g(z) = g_1(z) + g_2(z) = \sum_{n>0} \frac{a_{-n}}{z^n} + \sum_{n\geq 0} a_n z^n.$$

Set

$$u_1(z) = \text{Re } g_1(z), \quad u_2(z) = \text{Re } g_2(z).$$

Then, on Δ_a^* we have

$$u = u_1 + u_2 + \beta \log |z| + u(z_0).$$

The function u_1 is extended to a harmonic function on $\mathbf{C}^*$ and the limit of $u_1(z)$ exists as z tends to ∞. On the other hand, u_2 is extendable to a harmonic function on $\{z; |z| < a\}$ and so it is bounded on $\{z; |z| < a'\}$ for an arbitrarily fixed a' with $0 < a' < a$. Therefore, we can choose a positive constant A such that

$$ds^2 = |h|^2|dz|^2 \leq Ae^{2u_1(z)} \left(\frac{1}{|z|}\right)^{2(N+2)} |dz|^2 \tag{5.1.5}$$

on $\{z; 0 < |z| < a'\}$, where N is a positive integer with $N + 2 > -\beta$. Define a new metric

$$d\sigma^2 := Ae^{2u_1(z)} \left(\frac{1}{|z|}\right)^{2(N+2)} |dz|^2$$

on $\mathbf{C}^*$. By a change of variable $\zeta = 1/z$, we can rewrite

$$d\sigma^2 = Ae^{2u_1(1/\zeta)}|\zeta|^{2N}|d\zeta|^2$$

around $z = \infty$. Consider the holomorphic function

$$w(\zeta) := \int_0^\zeta \zeta^N e^{g_1(\zeta)} d\zeta,$$

on $\mathbf{C}$. Let R_0 be the least upper bound of all R satisfying the condition that there is a domain D such that $w(\zeta)$ maps D onto Δ_{R_0} properly, namely, $w^{-1}(K) \cap D$ is compact for any compact set K in Δ_{R_0}. The restriction of the map w to D gives a covering surface over Δ_{R_0}, which is ramified with ramified index N at the origin and unramified elsewhere. By the definition of R_0, we can find a smooth curve γ in D converging to the origin in the ζ-plane such that the image of $w \circ \gamma$ is a segment from the origin to the boundary of Δ_{R_0}. By the assumption, we see

$$\infty = \int_\gamma ds \leq \int_\gamma d\sigma = \int_{w \circ \gamma} |dw| = R_0.$$

In this situation, we can show easily that there is a covering isomorphism between the covering surface $w : D \to \mathbf{C}$ and $w^* : \mathbf{C} \to \mathbf{C}$ defined by $w^*(\zeta) = \zeta^{N+1}$. Since there is no proper subdomain of $\mathbf{C}$ which is biholomorphic with $\mathbf{C}$ itself, D coincides with the total plane $\mathbf{C}$. Moreover, every biholomorhic automorphism Φ of $\mathbf{C}$ which leaves the origin fixed is necessarily represented as $\Phi(\zeta) = C\zeta$ for a positive constant C. This implies that g_1 is a constant. By (5.1.5) we have Lemma 5.1.4.

PROOF OF THEOREM 5.1.3. Obviously, (ii) implies (iii) because of the representation (5.1.2). Assume that M is biholomorphic with an open subset of a compact Riemann surface $\bar{M}$ and the Gauss map $G : M \to P^{m-1}(\mathbf{C})$ is extended to a holomorphic map $\tilde{G} : \bar{M} \to P^{m-1}(\mathbf{C})$. We then have $G^*(\mathrm{Ric}\,[ds^2]) = \tilde{G}^*(\mathrm{Ric}\,[ds^2])$ on M. Therefore,

$$0 \geq C(M) = \int_M \mathrm{Ric}\,[ds^2] \geq \int_{\bar{M}} \mathrm{Ric}\,[ds^2] > -\infty,$$

which shows that the condition (i) is satisfied.

Now, we assume that M is of finite total curvature. By Theorem 4.2.7 M is of finite type, namely, there is a compact Riemann surface $\bar{M}$ such that M is biholomorphic with $\bar{M} - \{p_1, \cdots, p_k\}$. For each p_j we take a neighborhood U_j on which there is a holomorphic local coordinate z such that $z(p_j) = \infty$ and $U_j = \Delta_{s,\infty} := \{z; s < |z| \leq +\infty\}$. We represent the Gauss map G as $G = (f_1 : \cdots : f_m)$ with holomorphic functions $f_i := \partial x_i / \partial z$ $(1 \leq i \leq m)$ on $\Delta_{s,\infty}$. Then, the induced metric ds^2 is given by (1.2.8), which is rewritten as

$$ds^2 = (|f_1|^2 + \cdots + |f_m|^2)|dz|^2$$

for a reduced representation $G = (f_1 : \cdots : f_m)$. This implies that

$$-\mathrm{Ric}\,[ds^2] = \Omega_G, \tag{5.1.6}$$

where Ω_G denotes the pull-back of the Fubini-Study metric form on the projective space $P^{m-1}(\mathbf{C})$. Consider the order function

$$T_G(r) = \int_s^r \frac{dt}{t} \int_{\Delta_{s,t}} -\mathrm{Ric}\,[ds^2].$$

By the assumption, it satisfies the condition

$$\lim_{r\to\infty} \frac{T_G(r)}{\log r} < +\infty.$$

We apply here Proposition 3.3.3 to the Gauss map G. We can conclude G has a removable singularity at ∞. Take a new holomorphic local coordinate $\zeta := 1/z$ on U_j and choose a new reduced representation

$G = (g_1 : \cdots : g_m)$ around the origin. Here, by shrinking U_j and changing the indices $i = 1, 2, \ldots, m$ if necessary, we may assume that g_1 has no zero and $\nu_{g_1}(0) = \min_{1 \le i \le m} \nu_{g_i}(0)$. Then, each g_i/g_1 is holomorphic at $\zeta = 0$ and so bounded around the origin. We can find a positive constant C such that

$$ds^2 = 2|f_1|^2 \left(\sum_{i=1}^{m} \left| \frac{f_i}{f_1} \right|^2 \right) |d\zeta|^2 = 2|f_1|^2 \left(\sum_{i=1}^{m} \left| \frac{g_i}{g_1} \right|^2 \right) |d\zeta|^2 \le C|f_1|^2 |d\zeta|^2.$$

In this situation, we can apply Lemma 5.1.4 to the function $h := f_1$ to show that f_1 has at worst a pole at the origin. This implies that each ω_i is extendable to $\bar{M}$ as a meromorphic form. Thus, Theorem 5.1.3 is completely proved.

REMARK 5.1.7. Let $x : M \to \mathbf{R}^m$ be a complete minimal surface of finite total curvature immersed in $\mathbf{R}^m$. Then, it is easily seen that a compact Riemann surface $\bar{M}$ satisfying the condition (ii) of Theorem 5.1.3 is unique, namely, if we have two compact Riemann surfaces $\bar{M}_1$ and $\bar{M}_2$ such that there are biholomorphisms Φ_1 of M onto $\bar{M}_1 - \{p_1, \ldots, p_k\}$ and Φ_2 of M onto $\bar{M}_2 - \{q_1, \ldots, q_\ell\}$, then $k = \ell$ and the map $\Phi_2 \circ \Phi_1^{-1}$ is extended to a biholomorphism between $\bar{M}_1$ and $\bar{M}_2$. We shall call a Riemann surface $\bar{M}$ satisfying the condition of Theorem 5.1.3, (ii) a compactification of M in the following.

For a complete minimal surface $x : M \to \mathbf{R}^m$ of finite total curvature immersed in $\mathbf{R}^m$, take a compactification $\bar{M}$ of M. As a result of Theorem 5.1.3, the holomorphic forms $\omega_i := \dfrac{\partial x_i}{\partial z} dz$ $(1 \le i \le m)$ on M are extended as meromorphic forms on $\bar{M}$. The metric $ds^2 = 2(|\omega_1|^2 + \cdots + |\omega_m|^2)$ is also extended to a pseudo-metric on $\bar{M}$. We can consider the divisor

$$\nu_{ds} = \min(\nu_{\omega_1}, \ldots, \nu_{\omega_m})$$

on $\bar{M}$, where we define $\nu_{\omega_i} = \nu_{f_i}$ for $\omega_i = f_i dz$.

PROPOSITION 5.1.8. *Let $\bar{M}$ be a compactification of a nonflat complete minimal surface M immersed in $\mathbf{R}^m$ with finite total curvature and $\{p_1, \ldots, p_k\} := \bar{M} - M$. For each $\ell = 1, \ldots, k$, it holds that*

$$\nu_{ds}(p_\ell) \le -2.$$

PROOF. For each $\ell = 1, \ldots, k$ set $\nu_\ell := \nu_{ds}(p_\ell)$. Taking a holomorphic local coordinate z with $z(p_\ell) = 0$, we can write

$$ds^2 = |z|^{2\nu_\ell} h^2 |dz|^2$$

with a positive C^∞ function h. Our purpose is to show that some ν_ℓ is smaller than -1. Assume that $\nu_\ell \geq -1$ for all ℓ. Expand the functions $f_i := \partial x_i/\partial z$ $(1 \leq i \leq m)$ as

$$f_i(z) = \frac{c^i_{-1}}{z} + \sum_{n=0}^{\infty} c^i_n z^n,$$

We then have

$$\begin{aligned} x_i(z) &= 2\,\mathrm{Re}\left(\int_{z_0}^{z} f_i(z)dz\right) + x_i(z_0) \\ &= 2\,\mathrm{Re}\left(c^i_{-1}\log z + \sum_{n=0}^{\infty} \frac{c^i_n}{n+1} z^{n+1}\right) + x_i(z_0). \end{aligned}$$

The terms Re $(c^i_{-1} \log z)$ must be single-valued because all other terms are single-valued, which implies that the c^i_{-1} are real. On the other hand, we can write

$$\begin{aligned} 0 &= f_1^2 + \cdots + f_m^2 \\ &= \frac{(c^1_{-1})^2 + \cdots + (c^m_{-1})^2}{z^2} + \frac{d_{-1}}{z} + \sum_{n=0}^{\infty} d_n z^n \end{aligned}$$

with some constants d_n. It follows that

$$(c^1_{-1})^2 + \cdots + (c^m_{-1})^2 = 0.$$

Since the c^i_{-1} are all real, we have necessarily $c^i_{-1} = 0$ for $i = 1, \ldots, m$. This is impossible because M is complete with respect to ds^2. Thus we conclude Proposition 5.1.8.

To state another result on complete minimal surfaces of finite total curvature given by S. S. Chern and R. Osserman, we recall some properties of compact Riemann surfaces.

Let M be a Riemann surface. We denote the Euler characteristic of M by $\chi(M)$. As is well-known, if M is biholomorphic to a compact Riemann surface $\bar{M}$ with k points excluded, then $\chi(M) = 2 - 2\gamma - k$, where γ denotes the genus of $\bar{M}$.

Next, we consider a nonconstant holomorphic map of a compact Riemann surface $\bar{M}$ into $P^n(\mathbf{C})$. Take a hyperplane H in $P^n(\mathbf{C})$ with

$f(M) \not\subseteq H$. Then, $\sum_{z \in \bar{M}} \nu(f,H)(z)$ does not depend on a choice of H. In fact, for two hyperplanes

$$H_i : a_{i0}w_0 + \cdots + a_{in}w_n = 0$$

with $f(\bar{M}) \not\subseteq H_i$ $(i = 1,2)$, if we take a reduced representation $f = (f_1 : \cdots : f_n)$ and define a meromorphic function

$$\varphi := \frac{a_{20}f_0 + \cdots + a_{2n}f_n}{a_{10}f_0 + \cdots + a_{1n}f_n},$$

we have

$$\sum_{z \in \bar{M}} \nu(f,H_2)(z) - \sum_{z \in \bar{M}} \nu(f,H_1)(z) = \sum_{z \in \bar{M}} \nu_\varphi(z) = 0.$$

DEFINITION 5.1.9. We define the *degree* of $f : \bar{M} \to P^n(\mathbf{C})$ by

$$\deg(f) := \sum_{z \in \bar{M}} \nu(f,H)(z).$$

PROPOSITION 5.1.10. *Let f be a nonconstant holomorphic map of a compact Riemann surface $\bar{M}$ into $P^n(\mathbf{C})$ and denote by Ω_f the pull-back of the Fubini-Study metric on $\bar{M}$. Then*

$$\int_{\bar{M}} \Omega_f = \deg(f).$$

PROOF. For our purpose, we may assume that f is nondegenerate because we may replace $P^n(\mathbf{C})$ by the smallest projective linear subspace of $P^n(\mathbf{C})$ which includes $f(\bar{M})$. Consider a hyperplane

$$H_0 : w_0 = 0$$

and set

$$\{p_1, p_2, \ldots, p_k\} := f^{-1}(H_0).$$

Taking a reduced representation $f = (f_0 : \cdots : f_n)$ on a noncompact open set in M, we define the function

$$\psi := \log\left(1 + \left|\frac{f_1}{f_0}\right|^2 + \cdots + \left|\frac{f_n}{f_0}\right|^2\right),$$

which does not depend on the choices of the reduced representation of f and so ψ is well-defined on $\bar{M} - f^{-1}(H_0)$. For $\ell = 1, \ldots, k$ choosing holomorphic local coordinates z_ℓ with $z_\ell(p_\ell) = 0$ in a neighborhood of p_ℓ, we consider a sufficiently small positive ε such that $\bar{U}_\ell(\varepsilon) := \{z_\ell; |z_\ell| \leq \varepsilon\}$ are mutually disjoint. Since $dd^c \log |f_0|^2 = 0$ outside $f^{-1}(H_0)$, we obtain

$$\begin{aligned}
\int_{\bar{M}} \Omega_f &= \lim_{\varepsilon \to 0} \int_{\bar{M} - \cup_\ell U_\ell(\varepsilon)} \Omega_f \\
&= \lim_{\varepsilon \to 0} \int_{\bar{M} - \cup_\ell U_\ell(\varepsilon)} dd^c \psi = - \lim_{\varepsilon \to 0} \sum_{\ell=1}^{k} \int_{\partial U_\ell(\varepsilon)} d^c \psi \\
&= - \lim_{\varepsilon \to 0} \sum_{\ell=1}^{k} \int_{\partial U_\ell(\varepsilon)} d^c \log |F|^2 + \lim_{\varepsilon \to 0} \sum_{\ell=1}^{k} \int_{\partial U_\ell(\varepsilon)} d^c \log |f_0|^2,
\end{aligned}$$

where

$$|F|^2 := |f_0|^2 + \cdots + |f_n|^2$$

for a reduced representation of $f = (f_0 : \cdots : f_n)$ on $\cup_\ell U_\ell(\varepsilon)$. As $\varepsilon \to 0$, the first term of the last line obviously converges to zero. On the other hand, the last term is evaluated as

$$\begin{aligned}
\sum_{\ell=1}^{k} \int_{\partial U_\ell(\varepsilon)} d^c \log |f_0|^2 &= \sum_{\ell=1}^{k} \frac{1}{2\pi\sqrt{-1}} \int_{|z_\ell| = \varepsilon} \frac{f_0'(z)}{f_0(z)} dz \\
&= \sum_{\ell=1}^{k} \nu_{f_0}(p_\ell) = \deg\,(f)
\end{aligned}$$

by the argument principle. This completes the proof of Proposition 5.1.10.

We can apply Proposition 5.1.10 to minimal surfaces to obtain the following:

THEOREM 5.1.11. *Let $x = (x_1, \ldots, x_m) : M \to \mathbf{R}^m$ be a complete minimal surface immersed in $\mathbf{R}^m$ of finite total curvature and let $\bar{M}$ be a compactification of M with finitely many points added. Then, for the holomorphic extension $\tilde{G}$ of the Gauss map G of M to $\bar{M}$ it holds that*

$$C(M) = -\deg\,(\tilde{G}).$$

In the case of $m = 3$, for the meromorphic extension $\tilde{g}$ of the classical Gauss map g of M to $\bar{M}$ we have

$$C(M) = -2 \deg\,(\tilde{g}).$$

PROOF. By (5.1.6) and Proposition 5.1.10, we have

$$-C(M) = -\int_M \text{Ric}\,[ds^2] = \int_{\bar{M}} \Omega_{\tilde{G}} = \deg \tilde{G}.$$

For the case $m = 3$, by the same argument as in the proof of Theorem 1.3.8 the induced metric is given by

$$ds^2 = (|g_1|^2 + |g_2|^2)^2 |\omega|^2$$

for a nowhere vanishing holomorphic form ω and a reduced representation $g = (g_1 : g_2)$. This gives that

$$-\text{Ric}\,[ds^2] = 2\ \Omega_g$$

and therefore

$$-C(M) = 2 \int_{\bar{M}} \Omega_{\tilde{g}} = 2 \deg \tilde{g}.$$

This concludes the proof of Theorem 5.1.11.

Now, we give the following:

DEFINITION 5.1.12. Let ω be a nonzero meromorphic form on a compact Riemann surface. The *degree* of ω is defined by

$$\deg\,(\omega) := \text{ord}\ \nu_\omega = \sum_{z \in \bar{M}} \nu_\omega(z),$$

where ν_ω denotes the divisor defined by $\nu_\omega := \nu_f$ for each local expression $\omega = f dz$ in terms of a holomorphic local coordinate z.

THEOREM 5.1.13. *Let $\bar{M}$ be a compact Riemann surface and ω a nonzero meromorphic form on $\bar{M}$. Then*

$$\deg\,(\omega) = 2\gamma - 2 = -\chi(\bar{M}).$$

We omit the proof. For example, refer to [16, Theorem 17.12].

THEOREM 5.1.14 ([12, Theorem 2]). *Let $x = (x_1, \ldots, x_m) : M \to \mathbf{R}^m$ be a complete minimal surface immersed in $\mathbf{R}^m$ of finite total curvature and assume that M is biholomorphic to a compact Riemann surface $\bar{M}$ with k points removed. Then*

$$C(M) \leq \chi(M) - k.$$

PROOF. We set $\{p_1, \dots, p_k\} := \bar{M} - M$. Consider the holomorphic forms $\omega_i = \partial x_i$ $(1 \leq i \leq m)$. We can easily find a nonzero vector $(a_1, \dots, a_m)$ such that, for the form $\omega := a_1\omega_1 + \cdots + a_m\omega_m$,

$$\nu_\omega(p_\ell) = \nu_{ds}(p_\ell) \qquad (1 \leq \ell \leq k)$$

and, on a neighborhood of each p_ℓ a reduced representation $\tilde{G} = (\tilde{f}_1 : \cdots : \tilde{f}_m)$ of the Gauss map $\tilde{G}$ extended to $\bar{M}$ satisfies the condition

$$a_1\tilde{f}_1(p_\ell) + \cdots + a_m\tilde{f}_m(p_\ell) \neq 0.$$

Then, for the hyperplane

$$H : a_1w_1 + \cdots + a_mw_m = 0$$

we have

$$\nu(\tilde{G}, H)(z) + \nu_{ds}(z) = \nu_\omega(z) \qquad \text{for all } z \in \bar{M}.$$

Since $\nu_{ds}(z) = 0$ on M, we obtain

$$\deg\,(\tilde{G}) + \sum_{\ell=1}^{k} \nu_{ds}(p_\ell) = \deg\,(\omega) = -\chi(\bar{M}) = -\chi(M) - k$$

with the use of Theorem 5.1.13. Finally, according to Proposition 5.1.8, we conclude

$$\begin{aligned} C(M) = -\deg\,(\tilde{G}) &= (\chi(M) + k) + \sum_{\ell=1}^{k} \nu_{ds}(p_\ell) \\ &\leq (\chi(M) + k) - 2k = \chi(M) - k. \end{aligned}$$

§5.2 The Gauss map of minimal surfaces of finite total curvature

In this section, we shall describe theorems given by R. Osserman and S. S. Chern on value distribution of the Gauss map of complete minimal surfaces in $\mathbf{R}^m$ which have finite total curvature. We first explain the result in [62] which concerns complete minimal surfaces in $\mathbf{R}^3$.

THEOREM 5.2.1. *Let $x : M \to \mathbf{R}^3$ be a nonflat complete minimal surface immersed in $\mathbf{R}^3$ which has finite total curvature. Then, the classical Gauss map of M can omit at most three distinct points.*

For the proof, we use the following Riemann-Hurwitz relation.

THEOREM 5.2.2. *Consider a nonconstant holomorphic map f of a compact Riemann surface M_1 onto another compact Riemann surface M_2. Let γ_1 and γ_2 be the genera of M_1 and M_2 respectively, let d be the degree of f, namely the number of elements of $f^{-1}(p)$ for each $p \in M_2$ counted with multiplicities and let v be the sum of all ramification indices of f. Then,*

$$2\gamma_1 - 2 = v + d(2\gamma_2 - 2).$$

For the proof, see [16, p. 140].

PROOF OF THEOREM 5.2.1. By Theorem 5.1.3 there is a compact Riemann surface $\bar{M}$ such that M is biholomorphic with $\bar{M} - \{p_1, \dots, p_k\}$ and g is meromorphically extended to $\bar{M}$. We consider g as a holomorphic map of $\bar{M}$ into $P^1(\mathbf{C})$. We denote the degree of g by d. Assume that g omits q distinct values $\alpha_1, \dots, \alpha_q$. We denote by v_0 and v the sum of ramification indices of all ramified points of g in $\cup_j g^{-1}(\alpha_j)$ and in the total surface $\bar{M}$ respectively. Obviously,

$$v_0 \leq v. \tag{5.2.3}$$

Since $g^{-1}(\{\alpha_1, \dots, \alpha_q\}) \subseteq \bar{M} - M$ by assumption, we have

$$qd - v_0 \leq k. \tag{5.2.4}$$

On the other hand, by Theorem 5.2.2

$$2\gamma - 2 = v - 2d, \tag{5.2.5}$$

where γ denotes the genus of $\bar{M}$. Moreover, with the help of Theorems 5.1.11 and 5.1.14 we have

$$C(M) = -2d \leq \chi(M) - k = \chi(\bar{M}) - 2k.$$

This gives

$$2(k - d) \leq 2 - 2\gamma = 2d - v \leq 2d - v_0 \leq 2d - qd + k$$

with the use of (5.2.5), (5.2.3) and (5.2.4). Therefore,

$$k \leq (4 - q)d.$$

Since $k > 0$, we conclude $q < 4$. This gives Theorem 5.2.1.

To describe a generalization of Theorem 5.2.1 to the case of minimal surface in $\mathbf{R}^m$, we recall Plücker's formula for a holomorphic curve in $P^n(\mathbf{C})$.

Let f be a nonconstant holomorphic map of a compact Riemann surface $\bar{M}$ into $P^n(\mathbf{C})$ and let $f^k : \bar{M} \to P^{N_k}(\mathbf{C})$ be the derived curves of f, where $N_k = \binom{n+1}{k+1}$.

We give here the following:

DEFINITION 5.2.6. Let μ_k be the number defined by (3.2.1) for the k-th derived curve of f. We call the order $\sum_{z \in \bar{M}} \mu_k(z)$ of μ_k *the ramification index* of order k for f and denote it by σ_k.

As in §3.1, taking a reduced representation $f = (f_0 : \cdots : f_n)$ of f and setting $F := (f_0, \dots, f_n)$, $F_k := F \wedge F' \wedge \dots \wedge F^{(k)}$ on a neighborhood of each point of $\bar{M}$, we consider the divisors $\nu_k := \nu_{|F_k|}$, which does not depend on a choice of a reduced representation of f and hence are defined on the totality of $\bar{M}$. By (3.2.1) and Proposition 2.2.11, we have

$$\mu_k = \nu_{k-1} - 2\nu_k + \nu_{k+1} \tag{5.2.7}$$

and so

$$\sigma_k = \sum_{z \in \bar{M}} \nu_{k-1}(z) - 2\nu_k(z) + \nu_{k+1}(z). \tag{5.2.8}$$

Plücker's formula is stated as follows :

THEOREM 5.2.9. *Let f be a nondegenerate holomorphic map of a compact Riemann surface $\bar{M}$ into $P^n(\mathbf{C})$. Then, for the degree d_k of f^k and the ramification index σ_k, it holds that*

$$\sigma_k + d_{k-1} - 2d_k + d_{k+1} = 2\gamma - 2, \qquad (0 \le k \le n-1),$$

where we set $d_{-1} = d_n = 0$.

PROOF. Take nonzero decomposable vectors $A^\ell \in \bigwedge^\ell \mathbf{C}^{n+1}$ for $\ell = k-1, k, k+1$ and define the meromorphic form

$$\xi := \frac{\langle F_{k-1}, A^{k-1} \rangle \langle F_{k+1}, A^{k+1} \rangle}{\langle F_k, A^k \rangle^2} dz.$$

This does not depend on the choices of reduced representation of f or holomorphic local coordinate z and so is well-defined on the totality of $\bar{M}$.

For, if we choose another reduced representation hF, then the denominator and numerator are both multiplied by the same function $h^{2(k+1)}$. Moreover, if z is replaced by another holomorphic local coordinate ζ, then the denominator and numerator are multiplied by $\left(\frac{dz}{d\zeta}\right)^{k(k+1)}$ and $\left(\frac{dz}{d\zeta}\right)^{k(k-1)/2+(k+1)(k+2)/2}$ respectively and so ξ is unchanged. Since

$$\sum_{z\in\bar{M}} \nu_{(F_\ell, A^\ell)}(z) = d_\ell + \sum_{z\in\bar{M}} \nu_\ell(z)$$

for each ℓ, with the use of Theorem 5.1.13 and (5.2.8) we can conclude the desired identity

$$\begin{aligned} 2\gamma - 2 &= \sum_{z\in\bar{M}} \nu_\xi(z) \\ &= d_{k-1} - 2d_k + d_{k+1} + \sum_{z\in\bar{M}} (\nu_{k+1}(z) - 2\nu_k(z) + \nu_{k+1}(z)) \\ &= d_{k-1} - 2d_k + d_{k+1} + \sigma_k. \end{aligned}$$

Now, we give the following generalization of Theorem 5.2.1 to minimal surfaces in $\mathbf{R}^m$, which were given by S. S. Chern and R. Osserman([12]) for the case where the Gauss map is nondegenerate and by M. Ru([64]) for the general case.

THEOREM 5.2.10. *Let $x : M \to \mathbf{R}^m$ be a nonflat complete minimal surface immersed in $\mathbf{R}^m$ and let $G : M \to P^{m-1}(\mathbf{C})$ be the Gauss map of M. If M has finite total curvature, then the Gauss map of M cannot omit $m(m+1)/2$ hyperplanes H in $P^{m-1}(\mathbf{C})$ with $f(M) \not\subseteq H$ which are located in general position.*

PROOF. Assume that the Gauss map G omits q hyperplanes $H_1, \ldots, H_q$ with $f(M) \not\subseteq H_j$ which are located in general position. By Theorem 5.1.3, M is biholomorphic with a compact Riemann surface $\bar{M}$ with k points $p_1, \ldots, p_k$ removed and the Gauss map G can be extended to a holomorphic map of $\bar{M}$ into $P^N(\mathbf{C})$, where $N = m-1$. Take the smallest projective linear subspace of $P^N(\mathbf{C})$ which includes $G(\bar{M})$ and denote it by $P^n(\mathbf{C})$. We regard $\tilde{H}_j := H_j \cap P^n(\mathbf{C})$ as hyperplanes in $P^n(\mathbf{C})$, which are located in N-subgeneral position, and G as a map into $P^n(\mathbf{C})$, which is nondegenerate. By $\omega(j)$ and θ we denote Nochka weights and a Nochka constant for these hyperplanes in $P^n(\mathbf{C})$, respectively.

Set $d_0 := \deg(G)$ and

$$m_{j\ell} := \nu(G, \tilde{H}_j)(p_\ell)$$

for $j := 1, \dots, q$ and $\ell = 1, \dots, k$. Since $f^{-1}(H_j) \subseteq \bar{M} - M$, we have

$$\sum_{\ell=1}^{k} m_{j\ell} = d_0.$$

For each ℓ we have $\#\{j; m_{j\ell} > 0\} \leq m$ because the H_j's are located in general position. Take a subset A_ℓ of $\{1, \dots, q\}$ such that $\#A_\ell = m$ and $m_{j\ell} = 0$ for all $j \notin A_\ell$. Then,

$$\sum_{j=1}^{q} \omega(j) m_{j\ell} = \sum_{j \in A_\ell} \omega(j) m_{j\ell} \qquad (1 \leq \ell \leq q).$$

Setting $E_j := e^{m_{j\ell}}$, we apply Proposition 2.4.15. We can find indices $j_0, \dots, j_n$ in A_ℓ such that $\tilde{H}_{j_0}, \dots, \tilde{H}_{j_n}$ are located in general position in $P^n(\mathbf{C})$ and

$$\prod_{j \in A_\ell} (e^{m_{j\ell}})^{\omega(j)} \leq \prod_{\lambda=0}^{n} e^{m_{j_\lambda \ell}},$$

namely,

$$\sum_{j \in A_\ell} \omega(j) m_{j\ell} \leq \sum_{\lambda=0}^{n} m_{j_\lambda \ell}. \tag{5.2.11}$$

According to Lemma 3.5.8, for each p_ℓ, taking a holomorphic local coordinate ζ with $\zeta(p_\ell) = 0$ in some neighborhood of p_ℓ and changing homogeneous coordinates on $P^n(\mathbf{C})$ suitably, we can represent the map G as $G = (g_0 : \cdots : g_n)$ with holomorphic functions

$$g_i(\zeta) = \zeta^{\delta_i} + \text{ terms of higher order} \qquad (0 \leq i \leq n), \tag{5.2.12}$$

where $\delta_0 := 0 < \delta_1 < \cdots < \delta_n$. Then, we have

$$\mu_k = \delta_{k+1} - \delta_k - 1. \tag{5.2.13}$$

In fact, by (5.2.7) and (3.5.11)

$$\mu_k = \nu_{k+1} - \nu_k - (\nu_k - \nu_{k-1}) = \delta_{k+1} - (k+1) - (\delta_k - k).$$

We claim that

$$\max\{m_{j_\lambda \ell}; 0 \leq \lambda \leq k\} \leq \delta_k \qquad (0 \leq k \leq n). \tag{5.2.14}$$

To this end, we may change the indices so that

$$m_{j_0 \ell} \leq \ldots \leq m_{j_n \ell}.$$

We assume that $m_{j_k \ell} > \delta_k$ for some k. Set

$$H_{j_\lambda} : a_{j_\lambda 0} w_0 + \cdots + a_{j_\lambda n} w_n = 0 \qquad (\lambda = 0, 1, \ldots, n)$$

and define

$$F_{j_\lambda} := a_{j_\lambda 0} g_0 + \cdots + a_{j_\lambda n} g_n = 0 \qquad (\lambda = 0, 1, \ldots, n).$$

By (5.2.12)

$$F_{j_\lambda} := a_{j_\lambda 0}(\zeta^{\delta_0} + \cdots) + \cdots + a_{j_\lambda n}(\zeta^{\delta_n} + \cdots) = 0 \qquad (\lambda = 0, 1, \ldots, n).$$

Expand both sides as power series in ζ and compare the coefficients of them. We easily obtain

$$a_{j_\lambda 0} = \cdots = a_{j_\lambda k} = 0 \qquad \text{for } \lambda = k, \ldots, n.$$

This implies that $F_{j_k}, \ldots, F_{j_n}$ are linearly dependent, which is a contradiction. Thus (5.2.14) is valid.

As a consequence of (5.2.14) and (5.2.11), we have

$$\sum_{j=1}^{q} \omega(j) m_{j\ell} \leq \delta_0 + \cdots + \delta_n.$$

On the other hand, by (5.2.13)

$$\delta_0 + \cdots + \delta_n = \sum_{k=0}^{n-1} (n-k)\mu_k + \frac{n(n+1)}{2}.$$

Moreover, by virtue of Theorem 5.2.9, it holds that

$$\begin{aligned}\sum_{z\in\bar{M}}\left(\sum_{k=0}^{n-1}(n-k)\mu_k(z)\right) &= \sum_{k=0}^{n-1}(n-k)\sigma_k \\ &= \sum_{k=0}^{n-1}(n-k)(2\gamma-2-d_{k+1}+2d_k-d_{k-1}) \\ &= \frac{n(n+1)}{2}(2\gamma-2)+(n+1)d_0.\end{aligned}$$

Consequently,

$$\begin{aligned}\left(\sum_{j=1}^{q}\omega(j)\right)d_0 &= \sum_{j=1}^{q}\sum_{\ell=1}^{k}\omega(j)m_{j\ell} \\ &= \sum_{\ell=1}^{k}\left(\sum_{j=1}^{q}\omega(j)m_{j\ell}\right) \\ &\leq \sum_{\ell=1}^{k}\left(\sum_{k=0}^{n-1}(n-k)\mu_k(p_\ell)+\frac{n(n+1)}{2}\right) \\ &\leq \sum_{z\in\bar{M}}(n-k)\mu_k(z)+\frac{kn(n+1)}{2} \\ &= \frac{n(n+1)}{2}((2\gamma-2)+k)+(n+1)d_0.\end{aligned}$$

We can apply here Theorem 5.1.14 to see

$$-d_0 = C(M) \leq (2-2\gamma-k)-k = 2-2\gamma-2k.$$

These imply

$$\left(\sum_{j=1}^{q}\omega(j)\right)d_0 \leq \frac{n(n+1)}{2}(d_0-k)+(n+1)d_0,$$

which we can rewrite as

$$\frac{kn(n+1)}{2} \leq \left(\frac{n(n+1)}{2}+n+1-\sum_{j=1}^{q}\omega(j)\right)d_0.$$

We recall Theorem 2.4.11 (ii). We obtain

$$0 < \frac{kn(n+1)}{2} \leq \left(\frac{n(n+1)}{2} - \theta(q - 2N + n - 1)\right) d_0.$$

Moreover, using Theorem 2.4.11 (iii), we conclude

$$\begin{aligned} q &< \frac{n(n+1)}{2}\frac{1}{\theta} + 2N - n + 1 \\ &\leq \frac{(2N-n+1)n}{2} + 2N - n + 1 \\ &= \frac{(N+1)(N+2) - (N-n)(N-n-1)}{2} \\ &\leq \frac{(N+1)(N+2)}{2} = \frac{m(m+1)}{2}. \end{aligned}$$

This completes the proof of Theorem 5.2.10.

§5.3 Modified defect relations for the Gauss map of minimal surfaces

In this section, we shall give some applications of the modified defect relations given in Chapter 4 to the Gauss map of complete minimal surfaces in $\mathbf{R}^m$.

We begin by explaining minimal surfaces with branch points in $\mathbf{R}^m$.

DEFINITION 5.3.1. Let M be an open Riemann surface and let

$$x = (x_1, \ldots, x_m) : M \to \mathbf{R}^m$$

be a nonconstant differentiable map. We call M a generalized minimal surface in $\mathbf{R}^m$ if each x_i is harmonic and the metric on M induced from $\mathbf{R}^m$ is conformal.

Taking a holomorphic local coordinate z, we set $\omega_i := \partial x_i$ and $\nu_i := \nu_{\omega_i}$ for $i = 1, \ldots, m$. By assumption, the ω_i are holomorphic forms on M. By S we denote the set of all points a such that $\nu_i(a) > 0$ for all i. Obviously, S is a discrete set in M. The map x is regular at a, since the Jacobian matrix of x at a is of rank two if and only if a is not contained in S. Points in S are called *branch points* of M. We consider the continuous pseudo-metric ds^2 locally defined by (1.1.13) on M. As is seen from

Proposition 1.2.4, $M - S$ is regarded as a minimal surface immersed in $\mathbf{R}^m$. This is the reason why M is called a generalized minimal surface. In the following sections, minimal surfaces will mean generalized minimal surfaces unless we specify otherwise.

DEFINITION 5.3.2. We call the divisor of the pseudo-metric ds^2, which is given by

$$\nu_{ds} = \min\{\nu_i; 1 \leq i \leq m\},$$

the *branching divisor* of M.

Take a nonzero holomorphic form ω_0 on M with $\nu_{\omega_0} = \nu_{ds}$. We can write the Gauss map G of $M - S$ as

$$G = (g_1 : g_2 : \cdots : g_m) \tag{5.3.3}$$

outside S with holomorphic functions g_i satisfying the identity $\omega_i = g_i\omega_0$ $(1 \leq i \leq m)$. The right hand side of this equation has a unique holomorphic extension across S. We define the Gauss map G of M as the holomorphic map of M into $P^{m-1}(\mathbf{C})$ which is given by the right hand side of (5.3.3).

DEFINITION 5.3.4. For an arbitrary compact subset K of M we set

$$\rho_{ds}^K := \inf\{\rho \geq 0; [\nu_{ds}] \prec \rho\Omega_G \text{ on } M - K\},$$

where we set $\rho_{ds}^K = +\infty$ if there is no ρ with the above property. We define the *branching order* of M by

$$\rho_{ds} := \inf\{\rho_{ds}^K; K \text{ is an arbitrary compact subset of } M\}.$$

Obviously, if $x : M \to \mathbf{R}^m$ has only finitely many branch points, or particularly if x is an immersion, then $\rho_{ds} = 0$.

THEOREM 5.3.5. *Let $x : M \to \mathbf{R}^m$ be a complete minimal surface and $G : M \to P^{m-1}(\mathbf{C})$ the Gauss map of M. Consider the smallest projective linear subspace $P^n(\mathbf{C})$ of $P^{m-1}(\mathbf{C})$ which includes $G(M)$. Chosse q hyperplanes $H_1, \dots, H_q$ in $P^{m-1}(\mathbf{C})$ such that $\tilde{H}_j = H_j \cap P^n(\mathbf{C})$ $(1 \leq j \leq q)$ are hyperplanes in $P^n(\mathbf{C})$ located in N-subgeneral position, where $N \geq n$ and $q > 2N - n + 1$. If M has infinite total curvature, then*

$$\sum_{j=1}^{q} \omega(j) D_G(\tilde{H}_j) \leq n + 1 + \frac{(1 + \rho_{ds})n(n+1)}{2},$$

where $\omega(j)$ are Nochka weights for $\tilde{H}_j$'s.

PROOF. Take a nonzero holomorphic form ω_0 with $\nu_{\omega_0} = \nu_{ds}$ and consider the functions g_i with $\partial x_i = g_i \omega_0$ for $i = 1, \dots, m$. Then, G has a reduced representation

$$G = (g_1 : \cdots : g_m)$$

on M and the metric of M is given by

$$ds^2 = 2(|\omega_1|^2 + \cdots + |\omega_m|^2) = 2(|g_1|^2 + \cdots + |g_m|^2)|\omega_0|^2.$$

Therefore, we have

$$-\mathrm{Ric}[ds^2] = \Omega_G + [\nu_{ds}], \tag{5.3.6}$$

where $\Omega_G = dd^c \log(|g_1|^2 + \cdots + |g_m|^2)$ is the pull-back of the Fubini-Study metric form on $P^{m-1}(\mathbf{C})$ by the Gauss map G. By the assumption, we can take a $m \times m$ unitary matrix (u_{ij}) such that, for the functions $\tilde{g}_{i-1} := \sum_{j=1}^m u_{ij} g_j (1 \leq i \leq m)$, $\tilde{g}_{n+1}, \dots, \tilde{g}_{m-1}$ are all identically zero. We regard G as a map of M into $P^n(\mathbf{C})$. Then G is nondegenerate as a map into $P^n(\mathbf{C})$ and the pull-backs of the Fubini-Study metric forms on $P^{m-1}(\mathbf{C})$ and on $P^n(\mathbf{C})$ are the same. Choose an arbitrary $\rho \geq 0$ such that $[\nu_{ds}] \prec \rho\Omega_G$ on M outside a compact set K. Then, by (5.3.6) we have

$$-\mathrm{Ric}[ds^2] \prec (\rho + 1)\Omega_G$$

on $M - K$. Taking the infimum of the right hand side for various ρ, we obtain

$$\rho_G \leq \rho_{ds} + 1.$$

Theorem 5.3.5 is now an immediate consequence of Corollary 4.3.12 and Theorem 5.1.3.

COROLLARY 5.3.7. *Let $x : M \to \mathbf{R}^m$ be a complete nonflat minimal surface with infinite total curvature and let $G : M \to P^{m-1}(\mathbf{C})$ be the Gauss map of M. If $G(M)$ is not included in any projective linear subspace of dimension less than n, then for arbitrary hyperplanes $H_1, \dots, H_q$ $(1 \leq j \leq q)$ in $P^{m-1}(\mathbf{C})$ with $G(M) \not\subseteq H_j$ and located in general position we have*

$$\sum_{j=1}^{q} D_G(H_j) \leq 2m - n - 1 + \frac{(1 + \rho_{ds})n(2m - n - 1)}{2}. \tag{5.3.8}$$

PROOF. Since $\rho_G \leq \rho_{ds} + 1$ as in the proof of Theorem 5.3.5, Corollary 5.3.7 is a direct result of Corollary 4.3.14.

COROLLARY 5.3.9. *Let M be a nonflat complete minimal surface immersed in $\mathbf{R}^m$ with infinite total curvature and let G be the Gauss map of M. Then, for arbitrary hyperplanes $H_1, \ldots, H_q$ in $P^{m-1}(\mathbf{C})$ located in general position,*

$$\sum_{j=1}^{q} D_G(H_j) \leq \frac{m(m+1)}{2}.$$

PROOF. By assumption, the Gauss map G is not a constant. We can apply Corollary 5.3.7 for some n with $1 \leq n \leq N$. Here, $\rho_{ds} = 0$ because M has no branch points. Therefore,

$$\begin{aligned}\sum_{j=1}^{q} D_G(H_j) &\leq 2m - n - 1 + \frac{n(2m-n-1)}{2} \\ &= \frac{m(m+1) - (m-n-1)(m-n-2)}{2} \\ &\leq \frac{m(m+1)}{2},\end{aligned}$$

which proves Corollary 5.3.9.

COROLLARY 5.3.10. *Let M be a nonflat complete minimal surface and let G be the Gauss map of M. If G omits q hyperplanes in general position, then $q \leq m(m+1)/2$.*

This is an immediate consequence of Corollary 5.3.9 for the case where M has infinite total curvature and of Theorem 5.2.10 for the other case.

Now, we consider a holomorphic curve in $\mathbf{C}^m$ given by a nonconstant holomorphic map $w = (w_1, w_2, \ldots, w_m) : M \to \mathbf{C}^m$. The space $\mathbf{C}^m$ is identified with $\mathbf{R}^{2m}$ by associating a point $(x_1 + \sqrt{-1}y_1, \ldots, x_m + \sqrt{-1}y_m) \in \mathbf{C}^m$ with $(x_1, y_1, \ldots, x_m, y_m)$. The curve $w : M \to \mathbf{C}^m$ is considered as a minimal surface $w = (x_1, y_1, \ldots, x_m, y_m) : M \to \mathbf{R}^{2m}$. By the Cauchy-Riemann's equations, we know

$$f_i = \frac{\partial x_i}{\partial z} = \sqrt{-1}\frac{\partial y_i}{\partial z} \qquad (1 \leq i \leq m).$$

Therefore, the Gauss map of M is given by

$$G = (f_1 : -\sqrt{-1}f_1 : \cdots : f_m : -\sqrt{-1}f_m)$$

and so $G(M)$ is included in the projective linear subspace

$$\{(u_1 : v_1 : \cdots : u_m : v_m) \in P^{2m-1}(\mathbf{C}); u_i = -\sqrt{-1}v_i \ (1 \leq i \leq m)\},$$

which we denote by $P^{m-1}(\mathbf{C})$. Particularly, if M is not included in any proper affine subspace of $\mathbf{C}^m$, G is nondegenerate as a map into $P^{m-1}(\mathbf{C})$. The Gauss map G considered as a map of M into $P^{m-1}(\mathbf{C})$ is the same as the complex Gauss map of M defined in [39, p. 369].

As an easy consequence of Theorem 5.3.5, we get the following improvement of [39, Theorem 2.10].

COROLLARY 5.3.11. *Let $w : M \to \mathbf{C}^m$ be a holomorphic curve in $\mathbf{C}^m$ which is complete and not included in any affine hyperplane and let G be the Gauss map of M considered as a map of M into the above-mentioned space $P^{m-1}(\mathbf{C})$. If M is not of finite type or else if G is transcendental, then*

$$\sum_{j=1}^{q} D_G(H_j) \leq m + \frac{(\rho_{ds} + 1)m(m-1)}{2}$$

for arbitrary hyperplanes $H_1, \ldots, H_q$ in $P^{m-1}(\mathbf{C})$ in general position.

§5.4 The Gauss map of complete minimal surfaces in $\mathbf{R}^3$ and $\mathbf{R}^4$

We consider the Gauss map of a nonflat minimal surface $x = (x_1, x_2, x_3) : M \to \mathbf{R}^3$. As is stated in §1.3, instead of the Gauss map we may study the classical Gauss map of M. The extended complex plane $\bar{\mathbf{C}}$ is identified with $P^1(\mathbf{C})$ by corresponding each point $z \in \bar{\mathbf{C}}$ to $(1 : z) \in P^1(\mathbf{C})$ and ∞ to $(0 : 1)$. The classical Gauss map may be considered a map into $P^1(\mathbf{C})$.

PROPOSITION 5.4.1. *For the classical Gauss map $g : M \to P^1(\mathbf{C})$, it holds that*

$$\rho_g \leq \rho_{ds} + 2.$$

PROOF. We regard the surface M as an open Riemann surface with the conformal metric which is induced from the standard metric on $\mathbf{R}^3$. Set $\omega_i := \partial x_i (i = 1, 2, 3)$. Then the classical Gauss map $g : M \to P^1(\mathbf{C})$ is given by

$$g = (\omega_1 - \sqrt{-1}\omega_2 : \omega_3)$$

according to (1.3.6). Take a reduced representation $g = (g_1 : g_2)$. In view of (1.3.6) and (1.3.11) the induced metric is given by

$$ds^2 = (|g_1|^2 + |g_2|^2)^2 |\omega|^2$$

with a nonzero meromorphic form ω. Since ds^2 is continuous and $|g_1|^2 + |g_2|^2$ vanishes nowhere, we see that ω has no pole and $\nu_\omega = \nu_{ds}$, so that

$$-\operatorname{Ric}[ds^2] = [\nu_{ds}] + 2\Omega_g \prec (\rho + 2)\Omega_g$$

whenever $[\nu_{ds}] \prec \rho\Omega_g$. This yields that

$$\rho_g \leq \rho_{ds} + 2,$$

which gives Proposition 5.4.1.

The Gauss map of a surface in $\mathbf{R}^3$ is nondegenerate if and only if M is nonflat, or equivalently M is not a plane. On the other hand, with each point $\alpha := (a_0 : a_1) \in P^1(\mathbf{C})$ by associating the hyperplane

$$H : a_1 w_0 - a_0 w_1 = 0$$

in $P^1(\mathbf{C})$, we may identify the set of all hyperplanes in $P^1(\mathbf{C})$ with $P^1(\mathbf{C})$ itself. Then, arbitrary points $\alpha_1, \ldots, \alpha_q \in P^1(\mathbf{C})$ are in general position if and only if they are mutually distinct. Therefore, Corollary 4.3.16 implies the following result.

THEOREM 5.4.2. *Let $x : M \to \mathbf{R}^3$ be a nonflat complete minimal surface which has infinite total curvature and let $g : M \to P^1(\mathbf{C})$ be the classical Gauss map. Then, for arbitrary distinct points $\alpha_1, \ldots, \alpha_q$ in $P^1(\mathbf{C})$,*

$$\sum_{j=1}^{q} D_g(\alpha_j) \leq 4 + \rho_{ds}. \tag{5.4.3}$$

By the same reason as in the proof of Corollary 5.3.10, we conclude the following result of X. Mo and R. Osserman([52]), which is an improvement of the main result of [36].

COROLLARY 5.4.4. *Let M be a nonflat complete minimal surface with infinite total curvature immersed in $\mathbf{R}^m$. Then there are at most four distinct points in $P^1(\mathbf{C})$ whose inverse images under the Gauss map $g : M \to P^1(\mathbf{C})$ are finite.*

Combining Corollary 5.4.4 with Theorem 5.2.1, we have the following theorem ([36]).

COROLLARY 5.4.5. *Let M be a nonflat complete minimal surface. Then the classical Gauss map of M can omit at most four distinct points.*

We have also the following result of X. Mo and R. Osserman.

COROLLARY 5.4.6. *Let M be a nonflat complete minimal surface. If the classical Gauss map of M omits four distinct values, then it takes all other values infinitely many times.*

PROOF. Assume that the Gauss map g of M omits four distinct values. Then, M has infinite total curvature by virtue of Theorem 5.2.1. On the other hand, In view of Corollary 5.4.4, g takes all other values infinitely many times.

In Corollary 5.4.5, the number four is the best-possible. There are many examples of nonflat complete minimal surfaces in $\mathbf{R}^3$ whose classical Gauss maps omit four values. In fact, for an arbitrary set E with $\#E \leq 4$, there exists a complete minimal surface immersed in $\mathbf{R}^3$ whose Gauss map omits precisely the set E. For the details, see [63, Theorem 8.3].

We give here an example of a minimal surface with infinite total curvature, for which the branching order is equal to two and equality holds for the modified defect relation (5.4.3).

Our construction uses the Enneper-Weierstrass representation theorem given as Theorem 1.3.12. We shall suitably choose meromorphic functions h and g on a simply connected open Riemann surface $\tilde{M}$ and show that, for the meromorphic 1-forms

$$\omega_1 = h(1-g^2)dz, \quad \omega_2 = \sqrt{-1}h(1+g^2)dz, \quad \omega_3 = 2hgdz,$$

the map

$$(5.4.7) \qquad (x_1, x_2, x_3) = \left(\operatorname{Re}\int_0^\zeta \omega_1(z), \operatorname{Re}\int_0^\zeta \omega_2(z), \operatorname{Re}\int_0^\zeta \omega_3(z)\right)$$

of $\tilde{M}$ into $\mathbf{R}^3$ gives a minimal surface with the desired properties.

To this end, taking nonzero distinct values a_1, a_2 in $\mathbf{C}$ and setting $a_3 := \infty$, we choose a meromorphic function ϕ on $\mathbf{C}$ such that ϕ always takes the values $a_i (1 \leq i \leq 3)$ with multiplicity three and all zeros of $\phi(z) - w$ are simple for every $w \in \mathbf{C} - \{a_1, a_2\}$. For the existence of such a function, see [44, p. 45]. Consider the domain $D := \mathbf{C} - (\phi^{-1}(a_1) \cup \phi^{-1}(a_2))$ and the analytic subset $V^* := \{(z, w); \phi(z) = w^3\}$ of $D \times \mathbf{C}$. Let M be the normalization of the closure of V^* in $D \times P^1(\mathbf{C})$ and let $\tilde{M}$ be the universal covering surface of $\hat{M}$. Then $\tilde{M}$ is canonically regarded as a covering surface over D with the projection $F : \tilde{M} \to D$.

Now, consider the multi-valued meromorphic function

$$\tilde{g}(\zeta) = (\phi \circ F)^{\frac{1}{3}}(\zeta)$$

on $\tilde{M}$. By the definition of $\tilde{M}$, $\tilde{g}$ has a single-valued branch on $\tilde{M}$, with which we define the function g. We next consider the multi-valued meromorphic function

$$\tilde{h}(\zeta) = \frac{F'(\zeta)}{((\phi \circ F)(\zeta) - a_1)^{\frac{1}{3}}((\phi \circ F)(\zeta) - a_2)^{\frac{1}{3}}}. \tag{5.4.8}$$

Since $(\phi \circ F) - a_1$ and $(\phi \circ F) - a_2$ have poles of order three at every point of $(\phi \circ F)^{-1}(\infty)$, $\tilde{h}$ has a single-valued branch on $\tilde{M}$, with which we define the function h. For these functions h and g, we shall prove that the minimal surface $x = (x_1, x_2, x_3) : \tilde{M} \to \mathbf{R}^3$ defined by (5.4.7) has all the desired properties.

In view of properties of g, the minimal surface $x : \tilde{M} \to \mathbf{R}^3$ has obviously infinite total curvature. Now, take a reduced representation $g = g_1/g_0$, or write $g = (g_0 : g_1)$ as a map into $P^1(\mathbf{C})$. The induced metric is given by

$$ds^2 = |h|^2(1 + |g|^2)^2 |d\zeta|^2 = \frac{|h|^2}{|g_0|^4}(|g_0|^2 + |g_1|^2)^2 |d\zeta|^2.$$

Obviously, $\nu_{ds} = 0$ at each point where g has no zero or no pole. The same is also true at each point where g has a pole because g_0 has a zero of order one and the denominator of the right hand side of (5.4.8) has a pole of order two at such a point. Moreover, $\nu_{ds}(\zeta) = 2$ whenever $g(\zeta) = 0$, because $\nu_{F'}(\zeta) = 2$. These show that $\nu_{ds} = 2\nu_{g_1}$. If we consider a bounded function $k = |g_1|^2/(|g_0|^2 + |g_1|^2) (\leq 1)$, we obtain

$$[\nu_{ds}] = 2dd^c \log |g_1|^2 = 2dd^c \log(|g_0|^2 + |g_1|^2) + dd^c \log k^2.$$

This gives $\rho_{ds} \leq 2$.

We next show that $\tilde{M}$ is complete. Take an arbitrary curve Γ in $\tilde{M}$ which tends to the boundary. For the curve $\Gamma' := F(\Gamma)$ in D we have

$$\begin{aligned} L = L_{ds}(\Gamma) &= \int_{\Gamma} \frac{|F'(\zeta)|(1 + |(\phi \circ F)(\zeta)|^{\frac{2}{3}})}{|(\phi \circ F)(\zeta) - a_1|^{\frac{1}{3}} |(\phi \circ F)(\zeta) - a_2|^{\frac{1}{3}}} |d\zeta| \\ &= \int_{\Gamma'} \frac{1 + |\phi(z)|^{\frac{2}{3}}}{|\phi(z) - a_1|^{\frac{1}{3}} |\phi(z) - a_2|^{\frac{1}{3}}} |dz|, \end{aligned}$$

and Γ' tends to the set $\partial D = \{\infty\} \cup \phi^{-1}(a_1) \cup \phi^{-1}(a_2)$ except the case $L = \infty$. On the other hand, there exists a positive constant C_0 such that

$$\frac{1 + |\phi(z)|^{\frac{2}{3}}}{|\phi(z) - a_1|^{\frac{1}{3}} |\phi(z) - a_2|^{\frac{1}{3}}} \geq C_0$$

for every $z \in D$. If Γ' tends to ∞, then

$$L \geq C_0 \int_{\Gamma'} |dz| = +\infty.$$

Otherwise, there exists some point $z_0 \in \phi^{-1}(a_1) \cup \phi^{-1}(a_2)$ such that Γ' tends to z_0. Changing indices if necessary, we may assume that $\phi(z_0) = a_1$. Then, we find a neighborhood U of z_0 on which

$$\frac{1 + |\phi(z)|^{\frac{2}{3}}}{|\phi(z) - a_1|^{\frac{1}{3}} |\phi(z) - a_2|^{\frac{1}{3}}} \geq C_1 \frac{1}{|z - z_0|}$$

because $\nu_{\phi - a_1}(z_0) = 3$. For the portion $\tilde{\Gamma}$ of Γ' in U, we obtain

$$L \geq \int_{\tilde{\Gamma}} \frac{1 + |\phi(z)|^{\frac{2}{3}}}{|\phi(z) - a_1|^{\frac{1}{3}} |\phi(z) - a_2|^{\frac{1}{3}}} |dz| \geq C_1 \int_{\tilde{\Gamma}} \frac{|dz|}{|z - z_0|} = +\infty.$$

This implies the completeness of $\tilde{M}$.

On the other hand, since $\phi(z) \neq a_1, a_2$ on D, we see

$$g(\zeta) \notin \{a_1^{1/3}, a_1^{1/3}\omega, a_1^{1/3}\omega^2, a_2^{1/3}, a_2^{1/3}\omega, a_2^{1/3}\omega^2\}$$

for every $\zeta \in \tilde{M}$, where ω denotes one of the primitive third roots of unity. Therefore, the modified defects of these values are all one. This shows that the inequality (5.4.3) is the best-possible in this case.

We next consider a complete minimal surface $x = (x_1, x_2, x_3, x_4) : M \to \mathbf{R}^4$ immersed in $\mathbf{R}^4$. We shall study the value distribution of the Gauss map G of M more precisely in this case. The map G is a map into $Q_2(\mathbf{C})$. We shall inquire into the structure of $Q_2(\mathbf{C})$. We consider the map ψ_1 into $P^1(\mathbf{C})$ defined by

$$\psi_1(w) = (w_1 - \sqrt{-1}w_2 : w_3 + \sqrt{-1}w_4)$$

for each $w = (w_1 : \ldots : w_4)$ in $Q_2(\mathbf{C}) - E$, where

$$E := \{(w_1 : \ldots : w_4) \in Q_2(\mathbf{C}); w_1 - \sqrt{-1}w_2 = w_3 + \sqrt{-1}w_4 = 0\}.$$

For each $u := (u_1 : \ldots : u_4) \in Q_2(\mathbf{C})$

$$\begin{aligned} \psi_1(u) &= ((u_1 - \sqrt{-1}u_2)(u_3 - \sqrt{-1}u_4) : u_3^2 + u_4^2) \\ &= ((u_1 - \sqrt{-1}u_2)(u_3 - \sqrt{-1}u_4) : -(u_1^2 + u_2^2)) \\ &= (u_3 - \sqrt{-1}u_4 : -(u_1 + \sqrt{-1}u_2)) \end{aligned}$$

except the case where these have no meaning. On the other hand, for $(w_1 : w_2 : w_3 : w_4) \in E$ we have $w_1 + \sqrt{-1}w_2 \neq 0$ or $w_3 - \sqrt{-1}w_4 \neq 0$. Therefore, for $w \in E$ the value $\lim_{u \notin E, u \to w} \psi_1(u)$ exists, by which we define $\psi_1(w)$. Similarly, for each $w := (w_1 : w_2 : w_3 : w_4) \in Q_2(\mathbf{C})$ we define

$$\psi_2(w) = \begin{cases} (w_1 - \sqrt{-1}w_2 : -w_3 + \sqrt{-1}w_4) & w = (w_1 : \cdots : w_4) \notin E' \\ \lim_{u \notin E', u \to w} \psi_2(u) & \text{otherwise.} \end{cases}$$

where $E' := \{w_1 - \sqrt{-1}w_2 = -w_3 + \sqrt{-1}w_4 = 0\}$. By the use of the maps ψ_1 and ψ_2, we define the map

$$\Psi := (\psi_1, \psi_2) : Q_2(\mathbf{C}) \to P^1(\mathbf{C}) \times P^1(\mathbf{C}).$$

The map Ψ is bijective. In fact, if we consider the map $\Psi^* : P^1(\mathbf{C}) \times P^1(\mathbf{C}) \to Q_2(\mathbf{C})$ defined by

$$\begin{aligned} &\Psi^*((z : w), (u : v)) \\ &\quad := (zu + wv : \sqrt{-1}(zu - wv) : wu - zv : -\sqrt{-1}(wu + zv)), \end{aligned}$$

we can check easily that $\Psi^* \circ \Psi$ and $\Psi \circ \Psi^*$ are both identity maps. Consequently, the quadric $Q_2(\mathbf{C})$ is biholomorphic with $P^1(\mathbf{C}) \times P^1(\mathbf{C})$. For a more precise description, refer to [46, §2].

As in the case of surfaces in $\mathbf{R}^3$, regarding M as a Riemann surface with a conformal metric and taking a holomorphic local coordinate z, we set $\omega_i = \partial x_i$ $(i = 1, 2, 3, 4)$ and define the map

$$g = (g_1, g_2) = ((\omega_1 - \sqrt{-1}\omega_2 : \omega_3 + \sqrt{-1}\omega_4), (\omega_1 - \sqrt{-1}\omega_2 : -\omega_3 + \sqrt{-1}\omega_4)).$$

Instead of the Gauss map $G : M \to Q_2(\mathbf{C})$ we consider the map $g : M \to P^1(\mathbf{C}) \times P^1(\mathbf{C})$, which we call the *classical Gauss map* of M in the following.

We can prove the following defect relation.

THEOREM 5.4.9. *Let $x : M \to \mathbf{R}^4$ be a complete minimal surface immersed in $\mathbf{R}^4$ which has infinite total curvature and let $g = (g_1, g_2) : M \to P^1(\mathbf{C}) \times P^1(\mathbf{C})$ be the classical Gauss map of M. Take mutually distinct q_1 points $\alpha_1, \ldots, \alpha_{q_1}$ and mutually distinct q_2 points $\beta_1, \ldots, \beta_{q_2}$ in $\bar{\mathbf{C}}$.*

(i) *If both g_1 and g_2 are nonconstant and*

$$\sum_{i=1}^{q_1} D_{g_1}(\alpha_i) > 2, \quad \sum_{j=1}^{q_2} D_{g_2}(\beta_j) > 2,$$

then

$$\frac{1}{\sum_{i=1}^{q_1} D_{g_1}(\alpha_i) - 2} + \frac{1}{\sum_{j=1}^{q_2} D_{g_2}(\beta_j) - 2} \geq 1.$$

(ii) *If g_2 is a constant and g_1 is nonconstant, then*

$$\sum_{j=1}^{q_1} D_{g_1}(\alpha_j) \leq 3.$$

PROOF. We first consider the situation as in (i). Take a reduced representation $g_k = (g_{k0} : g_{k1}) (= g_{k1}/g_{k0})$ for each $g_k : M \to P^1(\mathbf{C}) (k = 1, 2)$. Set

$$h = \frac{f_1 - \sqrt{-1} f_2}{g_{10} g_{20}},$$

where $f_i = \partial x_i / \partial z$. Then we have

$$\begin{aligned}
f_1 - \sqrt{-1} f_2 &= h g_{10} g_{20}, \\
f_3 + \sqrt{-1} f_4 &= (f_1 - \sqrt{-1} f_2) \frac{g_{11}}{g_{10}} = h g_{11} g_{20}, \\
f_3 - \sqrt{-1} f_4 &= -(f_1 - \sqrt{-1} f_2) \frac{g_{21}}{g_{20}} = -h g_{10} g_{21}, \\
f_1 + \sqrt{-1} f_2 &= \frac{f_1^2 + f_2^2}{f_1 - \sqrt{-1} f_2} = -\frac{f_3^2 + f_4^2}{f_1 - \sqrt{-1} f_2} = -h g_{11} g_{21}.
\end{aligned}$$

Therefore, the induced metric on M is given by

$$\begin{aligned}
ds^2 &= 2 \left(\sum_{i=1}^{4} |f_i|^2 \right) |dz|^2 \\
&= (|f_1 - \sqrt{-1} f_2|^2 + |f_1 + \sqrt{-1} f_2|^2 \\
&\qquad + |f_3 - \sqrt{-1} f_4|^2 + |f_3 + \sqrt{-1} f_4|^2) |dz|^2 \qquad (5.4.10) \\
&= |h|^2 (|g_{10}|^2 + |g_{11}|^2)(|g_{20}|^2 + |g_{21}|^2) |dz|^2.
\end{aligned}$$

Here, h is a nowhere zero holomorphic function by the assumption that $x : M \to \mathbf{R}^3$ is an immersion. By assumption, the maps g_1 and g_2 of M into $P^1(\mathbf{C})$ are both nondegenerate. Moreover, by (5.4.10), g satisfies the condition $(C)_{11}$ of Definition 4.3.9. Therefore, we can easily obtain the desired conclusion with the use of Theorem 4.3.10. If g_1 is not a constant and g_2 is a constant, then $\rho_{g_1} \leq 1$. The assertion (ii) is an immediate consequence of Theorem 4.3.10 for the case $L = 1$.

For other related results for minimal surfaces in $\mathbf{R}^4$, refer to the papers [36], [38] and [52].

§5.5 Examples

In this section we shall show that, for an arbitrary odd number m, the number $m(m+1)/2$ of Corollary 5.3.9 is the best-possible, namely, there exist complete minimal surfaces in $\mathbf{R}^m$ whose Gauss maps are non-degenerate and omit $m(m+1)/2$ hyperplanes in general position (cf., [35]).

The purpose of this section is to prove the following:

THEOREM 5.5.1. *For an arbitrarily given odd number m there is a complete minimal surface in $\mathbf{R}^m$ whose Gauss map is nondegenerate and omit $m(m+1)/2$ hyperplanes in $P^{m-1}(\mathbf{C})$ located in general position.*

To prove this, we first give the following algebraic lemma.

LEMMA 5.5.2. *Let m be an odd number. For $0 \leq t \leq (m-1)/2$ we consider $(t+1)m$ polynomials*

$$\begin{aligned} f_i(u) &= (u - a_0)^{m-i} && (1 \leq i \leq m) \\ f_{m+i}(u) &= (u - a_1)^{m-i}(u - b_1)^{i-1} && (1 \leq i \leq m) \\ &\cdots\cdots \\ f_{tm+i}(u) &= (u - a_t)^{m-i}(u - b_t)^{i-1} && (1 \leq i \leq m), \end{aligned}$$

where a_σ, b_τ are mutually distinct complex numbers. For suitably chosen a_σ and b_τ, m arbitrarily chosen polynomials among these polynomials are linearly independent.

PROOF. We first remark that the set of all points $(a_0, a_1, \ldots, a_t, b_1, \ldots, b_t)$ which do not satisfy the desired condition constitutes an algebraic set in $\mathbf{C}^{2t+1}$, and so is nowhere dense if it does not coincide with the total space. We shall prove Lemma 5.5.2 by induction on t. It is trivial for the case $t = 0$. Assume that Lemma 5.5.2 is valid for the case $\leq t-1$, where $t \geq 1$, so that there is a nowhere dense subset A of $\mathbf{C}^{2t-1}$ such that m arbitrarily chosen polynomials among $f_1, f_2, \ldots, f_{tm}$ are linearly independent if $(a_0, a_1, \ldots, a_{t-1}, b_1, \ldots, b_{t-1})$ is not contained in A. Consider polynomials f_i defined as in Lemma 5.5.2 which includes unknown constants a_t and b_t. Choose arbitrarily m polynomials $f_{i_0}, f_{i_1}, \ldots, f_{i_n}$ among f_j $(1 \leq j \leq q)$, where $n = m-1$ and $q = (t+1)m$. We have to show that $f_{i_0}, \ldots, f_{i_n}$ are linearly independent. For brevity, we set $i_{-1} := 0$ and $g_r := f_{i_r} (0 \leq r \leq n)$. By changing indices, we may assume that

$$i_0 < i_1 < \cdots < i_k \leq t(n+1) < i_{k+1} < \cdots < i_n$$

for some k. It suffices to consider the case where $k < n$ because of the induction assumption. We consider the Wronskian $W(g_0, \ldots, g_k)$ of $g_0, \ldots, g_k$, which does not vanish identically. Choose a point c with $W(g_0, \ldots, g_k)(c) \neq 0$. Replacing the coordinate u by $u + c$, we may assume that $c = 0$. Set

$$g_r(u) = \sum_{s=0}^{n} A_{rs} u^s \qquad (0 \leq r \leq n),$$

where A_{rs} may be considered as polynomials in a_σ and b_τ. Let $b_t := 0$. Then, $g_{k+1}, \ldots, g_n$ can be written as

$$g_r(u) = (u - a_t)^{\ell_r} u^{n-\ell_r} \qquad (k+1 \leq r \leq n)$$

for $\ell_r := (t+1)(n+1) - i_r$ and so

$$A_{rs} = (-1)^{n-s} \binom{\ell_r}{n-s} a_t^{n-s}$$

for $k+1 \leq r \leq n$ and $0 \leq s \leq n$, where we set $\binom{\ell}{s} = 0$ if $s < 0$. On the other hand, A_{rs} are independent of a_t for $0 \leq r \leq k$. It suffices to show that $F := \det(A_{rs}; 0 \leq r, s \leq n)$ does not vanish identically as a function of a_t. We apply the Laplace expansion theorem on determinants to the first $k+1$ rows and the last $n-k$ rows of $(A_{rs}; 0 \leq r, s \leq n)$. As is easily seen, F has no nonzero term of degree $< (n-k)(n-k-1)/2$ and the coefficient of the term of degree $(n-k)(n-k-1)/2$ of F with respect to a_t is given by

$$B = \det(A_{rs}; 0 \leq r, s \leq k) \det\left(\binom{\ell_r}{s-n+\ell_r}; k+1 \leq r, s \leq n\right).$$

The first and second terms equal $W(g_0, \dots, g_k)(0)$ and $\det(\ell_r^{n-s}; k+1 \leq r, s \leq n)$ respectively, up to a nonzero constant multiple. Therefore, we conclude $B \neq 0$. The proof of Lemma 5.5.2 is completed.

PROOF OF THEOREM 5.5.1. For a given odd number m we set $n := m-1$, $k := n/2$ and define m functions

$$h_{2\ell+1}(z) = e^{\ell z} + e^{(2k-\ell)z} \qquad (0 \leq \ell \leq k-1),$$

$$h_{2\ell+2}(z) = \sqrt{-1}(e^{\ell z} - e^{(2k-\ell)z}) \qquad (0 \leq \ell \leq k-1)$$

and

$$h_{2k+1}(z) = 2\sqrt{-k}e^{kz}.$$

Next, we take suitable constants a_σ $(0 \leq \sigma \leq k)$ and b_τ $(1 \leq \tau \leq k)$ such that the polynomials f_i $(1 \leq i \leq q := m(m+1)/2)$ have the properties in Lemma 5.5.2 for $t = k$. By changing the variable u suitably, we may assume that $a_0 = 0$. Set

$$M^* := \mathbf{C} - \{z; e^z = a_\tau \text{ or } e^z = b_\tau \text{ for some } \tau = 1, \dots, k\}$$

and consider the universal covering surface $\pi : M \to M^*$. Set

$$\psi(z) = \frac{1}{(e^z - a_1)(e^z - b_1)\cdots(e^z - a_k)(e^z - b_k)}$$

and define m holomorphic functions $\tilde{g}_i = \psi h_i$ $(1 \leq i \leq m)$ on M^*. Then we see easily

$$\tilde{g}_1^2 + \tilde{g}_2^2 + \cdots + \tilde{g}_m^2 = 0.$$

For brevity, we denote the functions $\tilde{g}_i \circ \pi$ and $\tilde{g}_i$ by the abbreviated notation g_i in the following.

We consider the functions x_i defined by (1.2.7) for the holomorphic forms $\omega_i := g_i dz$ $(1 \leq i \leq m)$. By Theorem 1.2.5 the surface $x = (x_1, \dots, x_m) : M \to \mathbf{R}^m$ is a minimal surface. Moreover, the Gauss map of M may be rewritten as $G = (g_1 : \cdots : g_m)$ and therefore as $G = (h_1 : \cdots : h_m)$. As is easily seen, a polynomial $P(u)$ vanishes identically if and only if $P(e^z)$ vanishes identically. Since the polynomials

$$P_{2\ell+1}(u) = u^\ell + u^{2k-\ell} \qquad (0 \leq \ell \leq k-1),$$

$$P_{2\ell+2}(u) = \sqrt{-1}(u^\ell - u^{2k-\ell}) \qquad (0 \leq \ell \leq k-1)$$

and

$$P_{2k+1}(u) = 2\sqrt{-k}u^k$$

are linearly independent over $\mathbf{C}$, the Gauss map of M is nondegenerate. Moreover, since $P_1, \ldots, P_m$ give a basis of the vector space of all polynomials of degree $\leq m-1$. we can find some constants c_{ij} such that

$$f_i = \sum_{j=1}^{m} c_{ij} P_j \qquad (1 \leq i \leq q).$$

Now, we consider q hyperplanes

$$H_i : c_{i1}w_1 + \cdots + c_{im}w_m = 0 \qquad (1 \leq i \leq q).$$

These are located in general position because m arbitrary polynomials among the f_i's are linearly independent. On the other hand, for each $j = 1, \ldots, q$ we can write

$$\sum_{j=1}^{m} c_{ij} h_j(z) = \sum_{j=1}^{m} c_{ij} P_j(e^z) = f_i(e^z) = (e^z - a_\tau)^{r_i}(e^z - b_\tau)^{s_i}$$

with suitable nonnegative integers r_i, s_i. In view of the definition of M^*, this implies that each $f_i(e^z)$ vanishes nowhere on M. Consequently, the Gauss map G of M omits q hyperplanes H_j located in general position.

To complete the proof of Theorem 5.5.1, it remains only to prove that the Riemann surface M with the induced metric ds^2 is complete. The metric ds^2 is given by (1.2.8). In our case, ds^2 is induced from the metric

$$d\tilde{s}^2 = 2v(z)^2|\psi(z)|^2|dz|^2$$

on M^* via the projection map of M onto M^*, where

$$\begin{aligned} v(z)^2 &= \sum_{\ell=0}^{k-1}(|e^{\ell z} + e^{(2k-\ell)z}|^2 + |e^{\ell z} - e^{(2k-\ell)z}|^2) + 4k|e^{kz}|^2 \\ &= 2\sum_{\ell=0}^{k-1}(|e^{\ell z}|^2 + |e^{(2k-\ell)z}|^2) + 4k|e^{kz}|^2. \end{aligned}$$

It suffices to prove that M^* is complete, because M is also complete in this case. For simplicity of notation, we denote the surface M^* by M and $d\tilde{s}^2$

by ds^2. We now take an arbitrary piecewise smooth curve $\gamma(t)(0 \leq t < 1)$ which tends to the boundary of M as t tends to 1. Our purpose is to show that the length of γ is infinite. The proof is given by reduction to absurdity. Assume that the length of γ is finite in the following.

We first consider the case where there exists a sequence $\{t_\nu\}$ with $\lim_{\nu\to\infty} t_\nu = 1$ such that $\{\gamma(t_\nu)\}$ has an accumulation point z_0 in $\mathbf{C}$. If $\gamma(t)$ does not tend to z_0 as t tends to 1, then the length of γ is obviously infinite. So, we see $\lim_{t\to 1}\gamma(t) = z_0$. Then, we have necessarily $e^{z_0} = a_\tau$ or $= b_\tau$ for some τ. Then we can write

$$e^z - e^{z_0} = (z - z_0)k(z)$$

with a holomorphic function k on a neighborhood of z_0 satisfying the condition $k(z_0) \neq 0$. Therefore, we can conclude

$$ds^2 \geq C_1^2|\psi(z)|^2|dz|^2 \geq C_2\frac{1}{|z - z_0|^2}|dz|^2$$

for positive constants C_1, C_2. This leads to a conclusion

$$L_{ds}(\gamma) = \int_\gamma ds \geq C_2\int_{\overline{z_1 z_0}}\frac{1}{|z - z_0|}|dz| = \infty,$$

where z_1 is a point sufficiently near z_0 and $\overline{z_1 z_0}$ denotes the line segment from z_1 to z_0. This contradicts the assumption.

Accordingly, we have only to study the case where $\gamma(t)$ tends to ∞ as t tends to 1. First, assume that $\{e^{\gamma(t)}\}$ is bounded. Then there is a positive constant C_3 such that $|v(z)\psi(z)| \geq C_3$ on the curve γ and so

$$L_{ds}(\gamma) = \int_\gamma ds \geq \sqrt{2}C_3\int_\gamma |dz| = \infty,$$

which contradicts the assumptions. Otherwise, there exists a sequence $\{t_\nu\}$ tending to 1 such that $\{e^{\gamma(t_\nu)}\}$ tends to ∞. Set $w = e^z$. Then $|dw| = |w||dz|$ and the metric is given by

$$\begin{aligned} ds^2 &= \frac{\sum_{\ell=0}^{k-1}(|w|^{2\ell} + |w|^{2(2k-\ell)}) + 4k|w|^{2k}}{|(w-a_1)(w-b_1)\cdots(w-a_k)(w-b_k)|^2}\frac{|dw|^2}{|w|^2} \\ &\geq \frac{4k}{|(1-a_1w^{-1})(1-b_1w^{-1})\cdots(1-a_kw^{-1})(1-b_kw^{-1})|^2}\frac{|dw|^2}{|w|^2}. \end{aligned}$$

Consider the curve

$$\gamma' : w(t) = e^{\gamma(t)},$$

which is divergent in the w-plane. We have

$$\int_{\gamma} ds \geq C_4 \int_{\gamma'} \frac{|dw|}{|w|} = \infty$$

for a positive constant C_4. Thus the proof of Theorem 5.5.1 is completed.

In case that the dimension m is even, we can obtain the same conclusion as in Theorem 5.5.1 for some particular cases. For an arbitrary even number m set $k := m/2$. In this case, we use entire functions

$$h_{2\ell+1} = e^{\ell z} + e^{(2k-\ell-1)z} \qquad (0 \leq \ell \leq k-2),$$

$$h_{2\ell+2} = \sqrt{-1}(e^{\ell z} - e^{(2k-\ell-1)z}) \qquad (0 \leq \ell \leq k-2),$$

and

$$h_{2k-1} = \sqrt{-k}(e^{(k-1)z} + e^{kz}), \quad h_{2k} = \sqrt{-k}(e^{(k-1)z} - e^{kz}).$$

Instead of Lemma 5.5.2 we use the following conjecture, which has not yet been justified for general cases but only for $m \leq 16$.

CONJECTURE. *Set $k := m/2$ for an arbitrarily given even number m. Then m arbitrarily chosen polynomials among the $3k$ polynomials*

$$\begin{aligned} g_i(u) &= u^{i-1} && (1 \leq i \leq k) \\ g_i(u) &= (u-1)^{i-1} && (k+1 \leq i \leq 2k) \\ g_i(u) &= u^{i-k-1}(u-1)^{m-i+1} && (2k+1 \leq i \leq 3k) \end{aligned}$$

are linearly independent.

If the above conjecture is true for an even number m, then we can show that there exist m distinct constants $a_1 := 0, b_1 := 1, a_2, b_2, \ldots, a_k, b_k$ such that, for further polynomials

$$g_{3k+m(\ell-1)+i}(u) = (u-a_\ell)^{m-i}(u-b_\ell)^{i-1} \quad (1 \leq \ell \leq k-1, 1 \leq i \leq m),$$

any m polynomials of $g_1, g_2, \ldots, g_q$ are linearly independent, where $q = 3k + m(k-1) = m(m+1)/2$.

Taking constants a_σ and b_τ satisfying the above condition, we consider the universal covering surface M of the set

$$M^* = \mathbf{C} - \{z; e^z = a_i \text{ or } e^z = b_i \text{ for some } i = 1, \ldots, k\}$$

and, using the function

$$\tilde{\psi}(z) = \frac{1}{(e^z - 1)(e^z - a_2)(e^z - b_2)\cdots(e^z - a_k)(e^z - b_k)}$$

we define m holomorphic functions

$$g_i = \tilde{\psi} h_i \qquad (1 \leq i \leq m)$$

on M^*. Then, in a similar manner as in the previous case where m is even, we can prove that for the function x_i defined by (1.2.7), the surface $x = (x_1, x_2, \ldots, x_m) : M \to \mathbf{R}^m$ is a complete minimal surface whose Gauss map omits $m(m+1)/2$ hyperplanes in general position.

In conclusion, if $m(\geq 3)$ is odd, or if m is an even number for which the above conjecture holds, then the number $m(m+1)/2$ of Corollary 5.3.9 is the best-possible.

Bibliography

[1] L. V. Ahlfors, The theory of meromorphic curves, Acta Soc. Sci. Fenn. Nova Ser. A, **3**, No. 4(1941).

[2] L. V. Ahlfors, An extension of Schwarz's lemma, Trans, A. M. S., **43**(1938), 359–364.

[3] J. L. Barbosa and M. do Carmo, On the size of a stable minimal surface in $\mathbf{R}^3$, Amer. J. Math., **98**(1976), 515 – 528.

[4] E. F. Beckenbach and R. Bellman, Inequalities, Springer, Berlin, 1965.

[5] S. Bernstein, Sur un théorème de géométrie et ses applications aux équations aux dérivées partielles du type elliptique, Comm. de la Soc. Math. de Kharkov(2éme sér.) **15**(1915–1917), 38–45.

[6] E. Borel, Sur les zéros des fonctions entières, Acta Math. **20**(1897), 357–396.

[7] H. Cartan, Sur les zéros des combinaisons linéaires de p fonctions holomorphes données, Mathematica, **7**(1933), 5–31.

[8] C. C. Chen, On the image of the generalized Gauss map of a complete minimal surface in $\mathbf{R}^4$, Pacific J. Math., **102**(1982), 9–14.

[9] W. Chen, Cartan's conjecture: Defect relation for meromorphic maps from parabolic manifold to projective space, Ph. D. dissertation, Notre Dame University, 1987.

[10] W. Chen, Defect relations for degenerate meromorphic maps, Trans, A.M.S., **319**(1990), 499–515

[11] S. S. Chern, An elementary proof of the existence of isothermal parameters on a surface, Proc. Amer. Math. Soc., **6**(1955), 771 – 782.

[12] S. S. Chern and R. Osserman, Complete minimal surfaces in euclidean n-space, J. Analyse Math., **19**(1967), 15–34.

[13] M. J. Cowen and P. A. Griffiths, Holomorphic curves and metrics of negative curvature, J. Analyse Math., **29**(1976), 93–153.

[14] R. S. Earp and H. Rosenberg, On values of the Gauss map of complete minimal surfaces in $\mathbf{R}^3$, Comment. Math. Helv., **63**(1988), 579–586.

[15] R. Finn, On a class of conformal metrics, with application to differential geometry in the large, Comm. Math. Helv., **40**(1965), 1–30.

[16] O. Forster, Lectures on Riemann surfaces, Berlin-Heidelberg-New York, Springer-Verlag, 1981.

[17] H. Fujimoto, Extensions of the big Picard's theorem, Tôhoku Math. J., **24**(1972), 415–422.

[18] H. Fujimoto, Families of holomorphic maps into the projective space omitting some hyperplanes, J. Math. Soc. Japan, **25**(1973), 235–249.

[19] H. Fujimoto, On families of meromorphic maps into the complex projective space, Nagoya Math. J., **54**(1974), 21–51.

[20] H. Fujimoto, On meromorphic maps into the complex projective space, J. Math. Soc. Japan, **26**(1974), 272–288.

[21] H. Fujimoto, The uniqueness problem of meromorphic maps into the complex projective space, Nagoya Math. J., **58**(1975), 1–23.

[22] H. Fujimoto, A uniqueness theorem of algebraically non-degenerate meromorphic maps into $P^N(\mathbf{C})$, Nagoya Math. J., **64**(1976), 117–147.

[23] H. Fujimoto, Remarks to the uniqueness problem of meromorphic maps into $P^N(\mathbf{C})$, I, Nagoya Math. J., **71**(1978), 13–24.

[24] H. Fujimoto, Remarks to the uniqueness problem of meromorphic maps into $P^N(\mathbf{C})$, II, Nagoya Math. J., **71**(1978), 25–41.

[25] H. Fujimoto, Remarks to the uniqueness problem of meromorphic maps into $P^N(\mathbf{C})$, III, Nagoya Math. J., **75**(1979), 71–85.

[26] H. Fujimoto, Remarks to the uniqueness problem of meromorphic maps into $P^N(\mathbf{C})$, IV, Nagoya Math. J., **83**(1981), 153–181.

[27] H. Fujimoto, On meromorphic maps into a compact complex manifold, J. Math. Soc. Japan, **34**(1982), 527–539.

[28] H. Fujimoto, The defect relations for the derived curves of a holomorphic curve in $P^n(\mathbf{C})$, Tôhoku Math. J., **34**(1982), 141–160.

[29] H. Fujimoto, On the Gauss map of a complete minimal surface in $\mathbf{R}^m$, J. Math. Soc. Japan, **35**(1983), 279–288.

[30] H. Fujimoto, Value distribution of the Gauss map of complete minimal surfaces in $\mathbf{R}^m$, J. Math. Soc. Japan, **35**(1983), 663–681.

[31] H. Fujimoto, A finiteness theorem of meromorphic maps into a compact normal complex space, Sci. Rep. Kanazawa Univ., **30**(1985), 15–25.

[32] H. Fujimoto, Non-integrated defect relation for meromorphic maps of complete Kähler manifolds into $P^{N_1}(\mathbf{C}) \times \cdots \times P^{N_k}(\mathbf{C})$, Japan. J. Math., **11**(1985), 233–264.

[33] H. Fujimoto, A unicity theorem for meromorphic maps of a complete Kähler manifold into $P^N(\mathbf{C})$, Tôhoku Math. J., **38**(1986), 327–341.

[34] H. Fujimoto, Finiteness of some families of meromorphic maps, Kodai Math. J. **11**(1988), 47–63.

[35] H. Fujimoto, Examples of complete minimal surfaces in $\mathbf{R}^m$ whose Gauss maps omit $m(m+1)/2$ hyperplanes in general position, Sci. Rep. Kanazawa Univ., **33**(1988), 37–43.

[36] H. Fujimoto, On the number of exceptional values of the Gauss map of minimal surfaces, J. Math. Soc. Japan, **40**(1988), 235–247.

[37] H. Fujimoto, On value distribution of Gauss maps of minimal surfaces in $\mathbf{R}^m$, Sûgaku (in Japanese), **40**(1988), 312–321.

[38] H. Fujimoto, Modified defect relations for the Gauss map of minimal surfaces, J. Differential Geometry, **29**(1989), 245–262.

[39] H. Fujimoto, Modified defect relations for the Gauss map of minimal surfaces, II, J. Differential Geometry **31**(1990), 365–385.

[40] H. Fujimoto, Modified defect relations for the Gauss map of minimal surfaces, III, Nagoya Math. J., **124**(1991), 13–40.

[41] H. Fujimoto, On the Gauss curvature of minimal surfaces, J. Math. Soc. Japan, **44**(1992), 427–439.

[42] M. L. Green, Some Picard theorems for holomorphic maps to algebraic varieties, Amer. J. Math., **97**(1975), 43–75.

[43] P. Griffiths and J. Harris, Principles of algebraic geometry, John Wiley & Sons, New York, 1978.

[44] W. K. Hayman, Meromorphic functions, Oxford Math. Monographs, Clar-endon Press, Oxford, 1964.

[45] E. Heinz, Über die Lösungen der Minimalflächengleichung, Nachr. Akad. Wiss. Göttingen(1952), 51–56.

[46] D. A. Hoffman and R. Osserman, The geometry of the generalized Gauss map, Memoirs Amer. Math. Soc. **236**, 1980.

[47] D. A. Hoffman and R. Osserman, The Gauss map of surfaces in $\mathbf{R}^3$ and $\mathbf{R}^4$, Proc. London Math. Soc., (3) **50**(1985), 27–56.

[48] E. Hopf, On an inequalilty for minimal surfaces $z = z(x, y)$, J. Rat. Mech. Analysis, **2**(1953), 519–522.

[49] A. Huber, On subharmonic functions and differential geometry in the large, Comment, Math, Helv. **32**(1957), 13–72.

[50] S. J. Kao, On values of Gauss maps of complete minimal surfaces on annular ends, Math. Ann., **291**(1991), 315–318.

[51] H. B. Lawson, Lectures on minimal submanifolds, Vol. **1**, Publish or Perish Inc., Berkeley, 1980.

[52] X. Mo and R. Osserman, On the Gauss map and total curvature of complete minimal surfaces and an extension of Fujimoto's theorem, J. Differential Geometry **31**(1990), 343–355.

[53] R. Narasimhan, Complex analysis in one variable, Birkhäuser, Boston, 1985.

[54] R. Nevanlinna, Einige Eindeutigkeitssätze in der Theorie der meromorphen Funktionnen, Acta Math., **48**(1926), 367–391.

[55] R. Nevanlinna, Le théorème de Picard-Borel et la théorie des fonctions méromorphes, Gauthier-Villars, Paris, 1929.

[56] J. C. C. Nitsche, On an estimate for the curvature of minimal surfaces $z = z(x, y)$, J. Math. Mech., **7**(1958), 767–769.

[57] E. I. Nochka, On the theory of meromorphic functions, Soviet Math. Dokl., **27**(2)(1983), 377–381.

[58] J. Noguchi and T. Ochiai, Geometric function theory in several complex variables, Transl. Math. Monographs, Vol. **80**, A. M. S., 1990.

[59] R. Osserman, An analogue of the Heinz-Hopf inequality, J. Math. Mech., **8**(1959), 383–385.

[60] R. Osserman, On the Gauss curvature of minimal surfaces, Trans. A. M. S., **96**(1960), 115–128.

[61] R. Osserman, Minimal surfaces in the large, Comm. Math. Helv., 35(1961), 65–76.

[62] R. Osserman, Global properties of minimal surfaces in E^3 and E^n, Ann. of Math., **80**(1964), 340–364.

[63] R. Osserman, A survey of minimal surfaces, 2nd edition, Dover Publ. Inc., New York, 1986.

[64] M. Ru, On the Gauss map of minimal surfaces immersed in $\mathbf{R}^n$, J. Differential Geometry, **34**(1991), 411–423.

[65] M. Ru, On complete minimal surfaces of finite total curvature, preprint.

[66] A. Ros, The Gauss map of minimal surfaces, preprint.

[67] I. Satake, Linear algebra, Marcel Dekker, Inc., New York, 1975.

[68] B. V. Shabat, Distribution of values of holomorphic mappings, Transl. Math. Monographs Vol. **61**, AMS, 1985.

[69] W. Stoll, Introduction to value distribution theory of meromorphic maps, lecture notes in Math. **950**(1982), 210–359, Springer-Verlag, Berlin-Heidel-berg-New York.

[70] W. Stoll, The Ahlfors-Weyl theory of meromorphic maps in parabolic manifolds, Lecture notes in Math., **981**(1983), 101–129, Springer-Verlag, Berlin-Heiderberg-New York-Tokyo.

[71] J. Suzuki and N. Toda, Some notes on the theory of holomorphic curves, Nagoya Math. J., **81**(1981), 79–89.

[72] M. Tsuji, Potential theory in modern function theory, Maruzen Tokyo, 1959.

[73] H. Weyl and J. Weyl, Meromorphic functions and analytic curves, Princeton Univ. Press, Princeton, New Jersey, 1943.

[74] B. White, Complete surfaces of finite total curvature, J. Differential Geometry, **26**(1987), 315–326.

[75] H. Wu, The equidistribution theory of holomorphic curves, Princeton Univ. Press, Princeton, New Jersey, 1970.

[76] F. Xavier, The Gauss map of a complete non-flat minimal surface cannot omit 7 points of the sphere, Ann. of Math., **113**(1981), 211–214. Erratum, Ann. of Math., (2)**115**(1982), 667.

[77] S. T. Yau, Some function-theoretic properties of complete Riemannian manifolds and their applications to geometry, Indiana Univ. Math. J., **25**(1976), 659–670.

Index

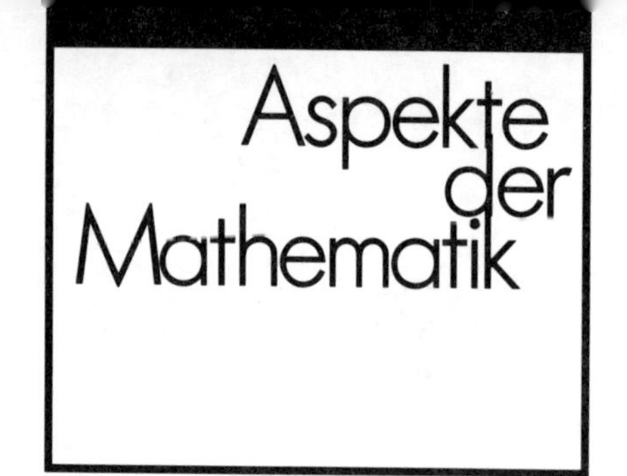

Edited by Klas Diederich

Band D 1: H. Kraft: Geometrische Methoden in der Invariantentheorie

Band D 2: J. Bingener: Lokale Modulräume in der analytischen Geometrie 1

Band D 3: J. Bingener: Lokale Modulräume in der analytischen Geometrie 2

Band D 4: G. Barthel/F. Hirzebruch/T. Höfer: Geradenkonfigurationen und Algebraische Flächen*

Band D 5: H. Stieber: Existenz semiuniverseller Deformationen in der komplexen Analysis

Band D 6: I. Kersten: Brauergruppen von Körpern

*A publication of the Max-Planck-Institut für Mathematik, Bonn